Jennifer Basterrechea

The
Great
Beach

The Great Beach

JOHN HAY

ILLUSTRATED BY DAVID GROSE

W · W · Norton & Company
New York—London

Library of Congress Cataloging in Publication Data

Hay, John, 1915
　The great beach.

　1. Natural history—Massachusetts—Cape Cod.
2. Cape Cod—Description and travel.　I. Title.
QH105.M4H36　1980　　917.44'9　　80–11105
ISBN 0–393–01367–7
ISBN 0–393–00983–1 pbk.

2 3 4 5 6 7 8 9 0

To Conrad and Mary Aiken
Henry and Gertrude Kittredge

Foreword

Since I first walked the beach, Cape Cod as a community has undergone some drastic changes. When I snatched at the local population figures as they were flying by nearly twenty years ago, the sum total for year-round residents was about 80,000. Now, in 1980, it is well over 140,000. Multiplying that by 3.5, the rule of thumb used by the Chamber of Commerce, you can get a rough idea of what the summer population amounts to. It was suggested in one grievous magazine article, partly disavowed later on, that in the face of what this was doing to the water supply in terms of its depletion and poisoning through waste material leached into the sandy soil, Cape Codders should immediately take to the hills. And there have been corresponding changes, such as a proliferation of bureaucratic regulations in every town, which would make some departed natives turn over in the sod and swear a little.

Having read my book, a young architectural student told me that he would like to walk the beach and spend a night or two on it for the purpose of a photographic essay. This was a few years after the National Seashore, by good fortune, was established by the Congress. He had no sooner walked a few hundred yards south of Highland Light in Truro, sleeping bag and equipment on his back, than he was picked up by the authorities and told that spending the night on the beach was no longer permitted, a change that I myself had not been aware of. So a little of an older freedom, or the space for it, had to be denied.

But the great beach is the center of greater changes besides which ours are lost in the tides. Sand, storm, and the great waters define it. Its forty-mile strip of white sands is only temporarily occupied by its human visitors. We are still just that and no more, visitors, obliged to move off at high tide, or forcibly kept from it during a northeast storm by its violent, elemental reactions. In spite of the beach buggies and campers that occupy the beach at various times of the year, it is hardly a highway in the common use of the term. We cannot really be said to own it. It is uninhabitable, still unconquered. Stand on top of its fragile cliffs and you have the whole sky around you, coming in with influences from all points of the compass. The open plains of the Atlantic roam with a wilderness light, and you are back with one of those high places on the continent that still offer you original space, the right scale and proportion for life on earth.

I suppose that anyone writing another book about Cape Cod could be accused of temerity in the face of such predecessors as the three Henrys—Thoreau, Beston, and Kittredge—as well as Dr. Wyman Richardson. However, each to his own eye. I have written about the Cape because of the circumstances of living there long enough to have begun to learn a little about it. The coast is long and the sea will not stop with the outer beach. All Americans who not only love nature but stand in awe of it will be more and more hard put to explain why we crowd our magnificent land and diminish it in proportion to the size of our demands. In *The Great Beach* I have given some of my own reasons.

FOREWORD

I am grateful to Dr. Alfred C. Redfield, Dr. John M. Zeigler, Mr. Joseph Chace, Dr. Loren C. Petry, Dr. Howard L. Sanders, and Dr. Ransom Somers for various assistance during the writing of this book, and hope they will not have any serious objections to the way I have used such information as they may have given me. This book also owes a great deal to the discerning and useful criticisms made by Richard K. Winslow of Doubleday.

Contents

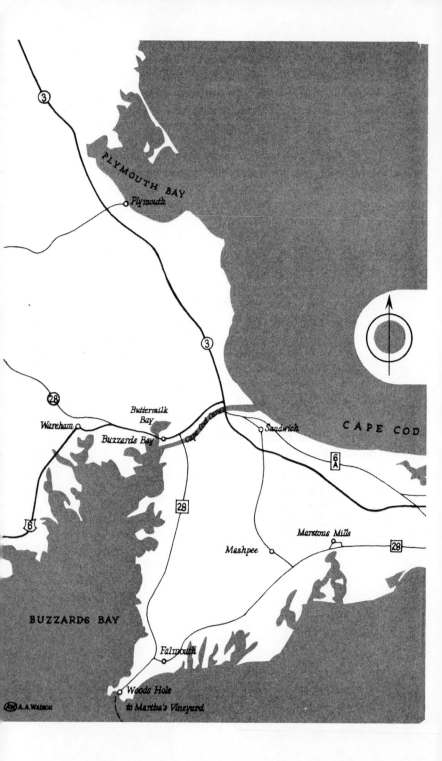

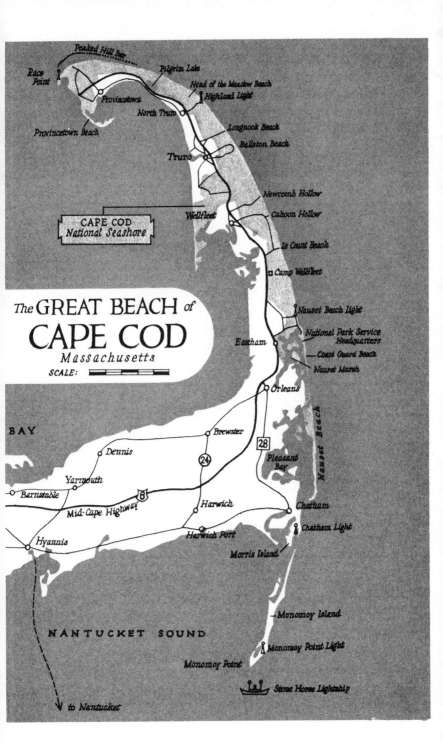

Peaked Hill Bar

Race Point

Pilgrim Lake

Provincetown

North Truro

Head of the Meadow Beach
Highland Light

Provincetown Beach

Longnook Beach
Ballston Beach

Truro

Newcomb Hollow

CAPE COD
National Seashore

Wellfleet

Cahoon Hollow

Le Count Beach

Camp Wellfleet

Nauset Beach Light

National Park Service
Headquarters
Coast Guard Beach
Nauset Marsh

The GREAT BEACH of
CAPE COD
Massachusetts
SCALE:

Eastham

Orleans

BAY

Brewster

28

Dennis

24

Pleasant
Bay

Yarmouth

Barnstable

6

Harwich

Nauset Beach

Mid-Cape Highway

Chatham

Hyannis

Harwich Port

Chatham Light

Morris Island

Monomoy Island

NANTUCKET SOUND

Monomoy Point Light

Monomoy Point

Stone Horse Lightship

to Nantucket

The
Great
Beach

Historical Distance

From a Distance

The Pilgrims who reached Cape Cod in 1620 had heard of it before. It got its name in 1602 and had been touched on by European seamen at least a century before that, and so when the Pilgrims ". . . fell in with that land which is called Cape Cod; the which being made and certainly known to be it, they were not a little joyful."

Their coming had taken a long time, and they had passed over "a tedious and dreadful" sea; but as Bradford's history relates it further: ". . . they now had no friends to welcome them, nor inns to entertain or refresh their weather-beaten bodies, no houses or much less towns to repair to, to seek for succor.

"And for ye season it was winter, and they that know ye winters of that country know them to be sharp and violent, and subject to cruell and fierce storms, dangerous to travel to known places, much more to search an unknown coast. Besides, what could they see but a hideous wilderness, full of wild beasts and wild men? and what multitudes there might be of them they knew not. Neither could they, as it were, go up to ye top of Pisgah, to view from this wilderness a more goodly country to feed their hopes; for which way soever they turned their eyes (save upward to the heavens) they could have little solace or content in respect of outward objects. For summer being done, all things stand upon them with a weather-beaten face; and the

alienation from environment —

whole country, full of woods and thickets, represented a wild and savage hue. If they looked behind them, there was the mighty ocean which they had passed, and was now as a main barrier and gulf to separate them from the civil parts of the world."

Now, nearly 350 years later, that lone lánd reaching out into the mighty ocean seems to be full of the "solace and content with respect to outward objects" which the Pilgrims lacked. Roads, gas stations, shopping centers, and a continually increasing number of houses, proclaim it as human territory, another populated home ground from which we have to go far to be separated from civilization. The simple, raw existence which the Pilgrims not only endured but anticipated has been replaced by a world of goods, which is not to say that we do not have to have a fortitude of our own, made inevitable, in great measure, by the very abundance we have achieved.

The Cape Cod of 1620 was more or less the same in its general outline as it is now, although the original woodland has been cut down, or burned over, to be replaced by less varied trees, much of the topsoil has eroded and blown away, and the shore line altered in the course of natural change. Superficially at least, it has been tamed, and in most areas the primal, unknown wildness is hard to imagine. The last thing you would expect to find on pulling in to a parking lot above a Cape Cod beach would be desolate wilderness, though if there is one, wilderness being in short supply these days, it would be well worth the effort to discover; but the sea, from which we are separated both by its vastness and the difference between water and air, could answer the description, and also the sands that define its limits.

Cape Cod's Outer Beach, stretching for forty miles from the tip of the Cape at Provincetown to the end of Monomoy Island is not undiscovered country. Many men have walked it. Planes skim over it in no time at all, and the beach buggies bruise it with impunity. Still, the marks we make on it are all erased in

human effect on G.B.

time. The sea and sand insist on their own art. The beach is in a continuous state of remaking and invites discovery. It was first called "Great" so far as I know, by Henry David Thoreau. Otherwise it has been known for a long time as the Outer Beach, the Outer Shore, or in more familiar terms as the Back Side. Now it forms a major part of the new National Seashore—in the process of establishment—and is therefore not owned by individuals, or the towns in which they reside, but by the people of the United States. It is under national protection and possession at the same time, so how we approach and treat its future is a very great responsibility, which is appropriate enough.

The beach, standing out against the sea, is a further limit to America before it shelves off into the Atlantic depths. For most travelers it means the end of a highway, a place of summer sands. It is in fact one end of a whole continent of roads, of communications, of the vast and intricate business of human passage. In a sense it used to be the other way around. With all the known parts of the civilized world behind them, the Pilgrims found in this beach not an end but a beginning, whatever it might entail, and that of course, is why they went there.

This is an age in which we are able to ignore or bypass the "tedious and dreadful" highway of the sea—city dwellers, road, rocket, car, and plane makers that we are—to the extent that we too may find it again for the first time. The beach, lying by the sea and sea invested, is always ready for a new kind of attention in a new world. That is the nature of the place. Cape Cod itself, now and ultimately, is at the disposition of the sea rather than human enterprise.

The Cape is a narrow peninsula, a little terminal arm jutting out in to the Atlantic, constructed of loose material left by the last glacier some 20,000 years ago. Its upper part, starting beyond High Head at Truro and forming the Provincetown hook, or hood, is of recent origin. It lacks the cliffs that stand over the beach from a mile or so north of Highland Light to as far as the Nauset Coast Guard Beach at Eastham, and for the most part has a history of deposition and accretion rather than removal. It was formed by storms, tides, and currents, piling in sand and other materials from the shore to the south, over bars and reefs of glacial debris. The sand is still packing up around Race Point, as it is also adding to the shore south of Nauset to the tip end of Monomoy Island, while storms take it away from other parts of the shore line. Within living memory a large island called Billingsgate, on which there was a lighthouse and at one time a "Try Yard" for whales, disappeared under the surface of Cape Cod Bay. It now appears as a shoal at low tide and is otherwise covered over by water, although the rocky lighthouse foundations still show above the surface in all but the highest tides. On the Bay side the shore line has been filled in in some areas, while it has receded in others, revealing for example, the bones of horses and cows in the bank at the head of a beach, which were once presumably, some distance behind it. Many a cottage owner after a storm has found his living-room floor with nothing below it but the tide.

Over the centuries great changes have occurred in the nature and extent of marshlands, inlets, ponds, estuaries, and beaches. No year, or even month, goes by without some alteration in the shore line. These changes, not always obvious, sometimes violent and immediate, are not such as to threaten the physical existence of Cape Cod for many thousands of years to come, but they are of the kind that accentuate its close relationship to water and tides and weather. As the map makers are well aware, it is not a static piece of land. It moves.

The trunk of the Cape starts out from the mainland and then that slender curving arm juts up and out into the water with a kind of brave assertion beyond the continental limits; but it is the shape and sweep of waves and sands, of molding and at the same time of pulling away that strikes you most about it, as if it were a conception to be made or discarded, standing out in its trial. The whole physical earth, in spite of its apparent constancies, its orbital speed, the speed of light, the regularity of the tides, the fine, exact balances to life, is subject to rhythmic change, or in a deeper sense, to re-creation.

From 20,000 feet up, Cape Cod looks very much as it does on topographic maps, its heights and depths eliminated, a flat level land of sandy margins and wide green patches emerging out of the sea. In fact, with all its glacial lakes and ponds—between three and four hundred in number—its streams, marshes, bays, coves, and inlets, it might seem to consist as much of water as of earth. On a clear day at a lower altitude, skirting or passing over the shore line, you can see configurations of sand, the slopes and curves of the shoals, the white swirls and scallops under water made by currents and tides. The sea sparkles, and explodes with light where the sun strikes it directly. The spilling waves make small white accents along the shore. Tilting in the heights, you get a sense of mobility on a great scale. All the close, pressing impressions of locality are replaced by the roving of the waters, the islands of the mapped world floating there, the height and weight and emptiness of the sky.

However far their ageless impunity may reach, the world's argument is that Cape Cod and its Outer Beach are under human guidance, surveillance, and authority. Those who come there bring their own distance with them. If we are not yet world-minded, we are world engaged. This is not a cast-off, self-sufficient countryside any longer, and it has lost most, if not all, of the look of a bleak, cut-over, and yet habitable seaside land that it had in the nineteenth century, when the inhabitants still depended on the sea for their livelihood, when you could smell the fish and hear the sermons on its shores.

An estimated 300,000 people visit the Cape during the summer, or even more, depending on the tides of economy and change, but after they have gone there are 80,000 year-round residents left, with more to be expected in the future. So, in spite of its stretches of comparatively uninhabited sands and its wooded areas, the Cape is caught up in the human scheme of things, and we can hardly avoid looking at it with modern eyes, for good or ill. We own it, and that is the way we are inclined to see it, not for its sake but ours. All roads lead to a Cape Cod beach, or to Los Angeles, or Yellowstone. Every place is invested with human importunity, and the crowd will tell you where you are.

Drive down any of the great concrete highways of the nation in the heat of the summer along with thousands, or millions, through a landscape whose scale affected our ideas of size to begin with, and you realize that Americans have an affinity for distance—which is also a capacity for laying the distance bare. We have learned this from our continent. We have learned how to exploit, turning the native, active riches of a great land into passive objects of our will, and we have taken a greatness from it for our own. While we have transformed our surroundings, we ourselves have been transformed without being altogether aware of the debt we owe.

Abstracted, in the summer months especially, to the terms of the contemporary world, some of Cape Cod's more crowded areas have a familiar, continental look. They are covered with asphalt, cars, motels, cheap housing, shops full of grotesque souvenirs with no relation to the place they serve, and they amount, when you come right down to it, to receiving grounds for power, made by a conquering civilization. Will it be the same on the moon? The great scale is in us, the effort and the risk of desolation.

The beach's openness is nearly filled with bodies, lying everywhere, or sitting, talking, absorbing the sun, or dashing suddenly into the relatively cold water, shouting, jumping, and splashing there, and then returning, flesh in warmth and radiance, performing the blessed ritual of doing nothing.

This hot surface, this wide open brilliance of sand, water, and sky is a summertime release for those in want. We claim it, and fill it with human demands; and yet it keeps its distance, resisting our bland assumption of authority.

Clouds like heaps of spun silk float up across the sky. The low waves splash along the sands, very lightly to the ear. Surfcasting rods are lodged in the sand, leaning out toward the water. Offshore, a white tern rises fluttering after a dive into the water, and a herring gull, large and deliberate by contrast, beats low over the surface. Behind all the crowd and the voices, hanging over like the intense and heavy sun, is a stillness, a suspension. Perhaps it is the soul of summer, that gives a provision of relief for those in want; and if we wait and watch there might be more to this beach and sea than what we came for. Waiting, in fact, seems to be its essence, since it gives no answer to what it is, being a wide, surface brightness, a tidal beat, a sounding whose monumental depths are concealed, suggesting too, that we might wait for it forever and know nothing.

nature upo a language
gives it a language
metaphor
refer nature to a manmade objects
human measurements

II

An Unimagined Frontier

One afternoon in the middle of June I set off from Race Point at Provincetown, carrying a pack and sleeping bag, with Nauset Light Beach in Eastham, twenty-five miles away, as my destination, and my purpose simply to be on the beach, to see it and feel it for whatever it turned out to be, since most of my previous visits had been of the sporadic hop, skip, and jump kind to which our automotivated lives seem to lead us.

The summer turmoil was not yet in full voice but the barkers were there on behalf of beach-buggy tours over the dunes, and a sight-seeing plane flew by; cars drew up and droned away, and families staggered up from the beach with their load of towels, shoes, bags, or portable radios. The beach did not contain quite the great wealth of paper, cans, bottles, and general garbage that it would later on, in July and August, but one of the first things to catch my eye as I lunged down on to the sands was an electric-light bulb floating in the water, a can of shaving soap, the remains of a rubber doll, and a great scattering of sliced onions—probably thrown off a fishing boat.

The air was dancing with heat. The sun seemed to have the power to glare through all things. With the exception of a camper's tent on the upper part of the beach, and a few isolated gray shacks perched on dune tops behind it, there was nothing ahead but the wide belt of sand curving around one unseen corner after another with the flat easing and stretching sea beside me. Two boys waved to me from where they were perched high up on a dune, and I waved back.

Then I heard an insistent, protesting bird note behind me, and a piping plover flew past. It was very pale, and sand colored, being a wild personification of the place it lived in. It suddenly volplaned down the slope of the beach ahead of me, fluttering, half disappearing in holes made by human feet, side-winged, edged away, still fluttering, in the direction of the shore line, and when it reached the water, satisfied, evidently, that it had led me far enough, it flew back. These birds nest on the beach above the high-tide line, and like a number of other species, try to lead intruders away when they come too close to their eggs and young.

With high, grating cries, terns flew over the beach and low over the water, occasionally plummeting in after fish. Among the larger species, principally common terns, there were some least terns—a tiny, dainty version of the "sea swallow," chasing each other back and forth. They have the graceful, sharply de-fined bodies and deep wingbeat of the other terns, but in their littleness and excitability they seem to show a kind of baby anger.

Also there were tree swallows gathering and perching on the hot, glittering sand, and on smooth gray driftwood just below the dunes. It was a band of them, adults, and young hatched during the early spring, chittering and shining with their bril-liant blue-green backs and white bellies.

It seemed to me that out of these birds—my unwilling or in-different companions—came a protest, the protest of a desert in its beauty, an ancient sea land claiming its rarity, with these rare inhabitants, each with its definition and assertion, each having the color and precision of life and place, out of an un-known depth of devising.

Behind the beach at Provincetown and Truro are eight square

miles of dunes, making a great series of dips and pockets, innumerable smooth scourings, hollows within wide hollows.
Standing below their rims are hills, mounds, and cones, chiseled
by the wind, sometimes flattened on the top like mesas. These
dunes give an effect of motion, rolling, dipping, roving, dropping down and curving up like sea surfaces offshore. When I
climbed the bank to see them I heard the clear, accomplished
notes of a song sparrow. There were banks of rugosa roses in
bloom, with white or pink flowers sending off a lovely scent, and
the dunes were patched with the new green of beach grass, bayberry, and beach plum, many of the shrubs looking clipped and
rounded, held down by wind and salt spray. The purple and
pink flowers of the beach pea, with purselike petals, were in
bloom too, contrasting with dusty miller with leaf surfaces like
felt, a soft, clear grayish-green. Down at the bottom of the hollows the light and wind catching heads of bunch grass, pinkish
and brown, waved continually; and the open sandy slopes were
swept as by a free hand with curving lines and striations.

A mile or so at sea, over the serene flatness of the waters, a
fishing boat moved very slowly by. I started down the beach
again, following another swallow that was twisting and dipping
in leafy flight along the upper edge of the beach. On the tide
line slippery green sea lettuce began to glimmer as if it had an
inner fire, reflecting the evening sun. I stopped somewhere a
mile or two north of Highland Light in Truro, built a small fire
of driftwood to heat up a can of food, and watched a bar appearing above the water as the tide ebbed. Low white waves conflicted
and ran across a dome of sand, occasionally bursting up like
hidden geysers.

The terns were still crying and diving as the sun's metal
light, slanting along the shore, began to turn a soft yellow, to
spread and bloom. They hurried back and forth, as if to make

use of the time left them, and fell sharply like stones into the shimmering road of light that led across the water.

Where I live on the lower Cape, that part of it which lies between the Cape Cod canal and Orleans, the land heads out directly to the sea, toward the east from the continental west. Cape Cod Bay lies to the north and Nantucket Sound to the south. The arm of the upper Cape turns in the Orleans area and heads up on a north-south axis, the head of it, or hand if you like, curving around so that the sandy barrens in the Provincetown area are oriented in an east to west direction again. I am used to looking toward Kansas to see the setting sun, and from the curving shore line at Truro I had the illusion that it was setting in the north and that when it rose the next morning it appeared to be located not very far from where it set, a matter of ninety or a hundred degrees. In fact it does set closer to the north at this time of year, and along the flat ocean horizon this becomes more clear to the eye, as well as its relative position at dawn and its arc during the day. On the open beach in spring and summer you are not only at the sun's mercy in a real sense, but you are also under wider skies. In the comparative isolation of the beach, which is convex, slanting steeply toward the water, and therefore hides its distances, I felt reoriented, turned out and around through no effort of my own, and faced in many possible directions.

Shortly before sundown a beach buggy, curtains at its windows and a dory attached, lumbered slowly down some preordained ruts in the sand, and then a smaller one passed by at the top of the low dunes behind me. Fishing poles were slung along the outside of both machines. It was getting to be a good time to cast for striped bass.

I sat on the sands and listened to the sonorous heave and splash of low waves. The sun, like a colossal red balloon filled with water, was sinking in to the horizon. It swelled, flattened, and disappeared with a final rapidity, leaving a foaming, fiery band behind it. I suddenly heard the wild, trembling cry of a loon behind me, and then saw it fly over, heading north. The wind grew cool, after a hot day when the light shone on metallic, glittering slow waters, and sharp, pointed beach grasses clicked together, while I watched the darkness falling around me.

A small seaplane flew by at low altitude, parallel to the shore. A sliver of a moon appeared and then a star; and then single lights began to shine on the horizon, while from the direction of Highland Light an arm of light shot up and swung around. A fishing boat passed slowly by with a light at its masthead and two—port and starboard—at its stern. A few night-flying moths fluttered near me. The sky began to be massive with its stars. I thought of night's legitimacies now appearing, the natural claim of all these single lights on darkness, and then, making my bed in a hollow just above the beach, I lowered down into infinity, waking up at about one o'clock in the morning to the sound of shouting, a strange direct interruption to the night. It was the loud implacable voice of the human animal, something very wild in itself, filling the emptiness.

"For Chrisake bring her higher up! I can't have her dig in that way." The tide had come in and someone was having trouble maneuvering his beach buggy along the thin strip of sand now available.

The light of dawn opened my eyes again before the sun showed red on the horizon, and I first saw the tiny drops of dew on tips and stems of beach grass that surrounded me. A sparrow sang, and then, somewhere behind the dunes, a prairie warbler with sweet notes on an ascending scale.

When I started walking again I caught sight of a young fox. Its fur was still soft and woolly and its gait had a cub's limpness where it moved along the upper edge of the beach. I wished the young one well, though I suspected it might have an uncomfort-

able life. In spite of an excessive population of rabbits, and their role in keeping it down, foxes have not been too highly regarded on the Cape. In recent years they seem to have been a skinny and somewhat delapidated bunch for the most part, suffering from parasitic skin diseases, and ticks in season. I once saw a fox out on an asphalt road sliding along on his chin and side, shoving and dragging himself in such a frantic way that I began to feel very itchy myself. I have heard them referred to in scornful way as "spoilers," fond of scavenging and rolling in dead meat. In other words, they are smelly, diseased and, to add another epithet "tricky," not to be trusted.

Yet this cub exploring an early morning on the sands had a future, however limited, and I remembered the lively trot of foxes when they are in good health, and their intelligence and curiosity, and simply their right to whatever special joys they might inherit.

I carried a pair of field glasses with me, along with the somewhat thoughtlessly assembled equipment I wore on my back and which seemed increasingly heavy as time went on. When not too conscious of my burden I would use the glasses to bring an inland or offshore bird closer to me. I noticed five eider ducks across the troughs of the waves, a remnant of the thousands that winter off the Cape along with such other sea birds as brant, Canada geese, scoters, mergansers, old squaws, and various members of the auk family. I passed a dead gannet lying on the sand. It had been badly oiled, reminding me of the hazards of jettisoned tanker or freighter oil to all these water birds which land on the sea to rest or feed.

There were a number of kingbirds on the dune rims, and they kept dropping down over the beach in their special way, to hover with fast wingbeat and flutter after flying insects. I heard the grating call of redwings, indicating marshy areas inland of the beach, but the cliffs above began to increase until they were

100 to 150 feet high or more, and the sun was so fierce that I had little interest in trying to scale them to see what was on the other side.

I plodded on, noticing very little after a while, my attention blunted, reduced to seeing that one foot got in front of the other. The more level upper parts of the beach provided fairly good walking, but the sand was soft, and to relieve my aching muscles I would then angle down to the water's edge where it was firmer, and there I was obliged to walk with one leg below the other because of the inclination of the beach. So I would return to the upper beach again and push ahead. I walked on, very hot and slow, seeing no one for miles until I came up to a group of bathers below a road and parking lot giving access to the beach, of the kind that are scattered along its reaches; and there I refilled my canteen at a cottage and went on.

I found that if I rested too long during this hike I had little desire to go on again, so I confined myself to an army "break" of ten minutes every hour. Renewed walking unlimbered me a little and the wind off the water cooled my sweating skin. I listened to the sound of the waves. In addition to their rhythmic plunge and splash, their breathing, they clashed occasionally with a sound like the breaking of heavy glass, the falling of timber, or a load of bricks.

I passed what was left of two shipwrecks during the day, a reminder of the dangers that still face ships along this coast with its fogs, its shifting winds, its storms, the hidden, treacherous offshore bars. The sands often reveal the timbers of old ships. One day their ribs, sodden and dark, barnacle encrusted, may reach up out of oblivion, and not long after that the water buries them under tons of sand. From them a local history calls out for recognition. Thousands of ships over three centuries wrecked on shoals, engulfed by violent seas, men with the dark of doom in them, to drown or to survive, and only a few timbers left to declare the ultimate dangers and their terror.

I was not in Death Valley, or on a raft at sea. My walk was not unusually long, and I could leave the beach if I had to, but the enormity of the area filled me more and more. It had so much in it that was without recourse. Its emptiness, the great tidal range beyond it and through it, the raw heartbeat of the waves, the implacable sun, established the kind of isolation and helplessness in me which the commerce and community of our lives tries so hard to disguise. Even the birds, I began to think, were more secure than I. They had their strong bright threads of cognizance to the areas they came to, the water, the sands, the marsh. They were fixed in entity and grace, eating what was theirs by evolution to be eaten, using land and air in the ways that had come to them, knowing this place and all places like it in terms of its bounds and boundlessness, meeting its naked eye in the ways they had been sent to do.

I started off in the morning admiring the brilliance of the sun, the small shadows from the dunes and across the beach, through driftwood, isolated beach plants and tidal wrack, with the wide flooding of light ahead and the variation in reflected light across the sea. I felt the sea moving quietly beside me. The waves heaved and sighed and spray was tossed lightly above the sand. Everything was continuous, untroubled, and deliberate; but as the day wore on the sun became my enemy, and I had very little rage or resource in me to fight it with. I was not fitted to enviromental stability, like a bird, or fox or fish. I found myself in an area of whose reaches I had never been wholly aware, and in me there was no mastery. The sun was not only hostile. It was an ultimate, an impossibility; and the waters beside me began to deepen from their pleasant daytime sparkle and freshness into an incalculable realm which I had hardly entered. I was touching on an unimagined frontier.

I spent my second night on the beach a few miles from Nauset Light where I left it the following morning. It was in the South Wellfleet area, and as I started to sleep on the sand a little above the high-tide line, I remembered that this was about the same place where a fishing boat had been wrecked two years before and two men drowned. I had seen the boat, with its cargo of fish, and some of the men's clothing strewn along the shore, and I had heard a little about the depths of their ordeal. Their story haunted me; and then I began to feel that I might be caught by the tide while I was asleep. There were only about twelve feet between the bottom of a steep cliff and the high-tide line. I would soon be lying on a narrow shelf at the sea's edge. So as the vague thought of being engulfed began to invade me, I took up my pack and sleeping bag again, retraced my steps down the beach, and found a way to the top of the cliff, where I spent the night in another hollow.

The light of dawn, lifting quickly out of the sea, flooding into the range of low-lying land, woke me up again, and it signaled to the birds, who started singing in all the thickets and heath around me with a sweet, high, shrill intensity, a kind of automatic worship; and after a while they quieted down again.

Little dirt roads dropped back from headlands through green slopes covered with bearberry and patches of yellow-flowered Hudsonia, or "poverty grass," and there were hollows dipping back inland, and woods of stunted pitch pine. From the top of the cliff I watched the sun starting to send light running across the blue table of the sea, making it glitter and move. The intensity of light and heat began to grow steadily as I walked down the beach again for the last stretch toward Nauset.

The beach is not so very far from where I live, or for that matter where anyone lives on the Cape. It is a few miles down the road, beyond the trees; and yet when I came back from my walk I felt as if I had been at enormous remove from my surroundings, caught out where I might have feared to be. The long line of sand and surf, the intensity of the sun, the cover of stars had come close enough to put me in council with that which had no answers. I was in awe of nature; and I understood that the sun and sea could be our implacable enemies. It was in this context that I saw our human world as subject to a stature that it never made.

III

The Resources of the Sea

Sit inland on the ground on a sunny day, and color, shadows, sound, substance, novelty in great detail, invade the smallest areas. One flower may attract many species of insects, brilliantly patterned and colored, flicking around, crawling, eating, gathering pollen, in any number of arresting ways, and the growth of plants around you, the shape of leaves, the general stir of things comes running like a carnival.

On the beach you might see a lone dragger lifting and falling, moving slowly parallel to the shore, beyond the measured fall of the green surf. A herring gull flies by. The vast sky swings overhead; the wind flies down the sand. Purple stones, driftwood, an occasional dead skate or dogfish comes to your attention as you walk on. A black crow pecks at seaweed far ahead. A sanderling flits by. You notice a finger sponge attached to a large mussel or a sea scallop, broken loose and washed in from offshore beds, and that seems to be all, in a relatively empty world; but between these single things, a grain of sand, a stone, a bird or bird track, a wave, you become conscious of a bounty of space.

The sea and its shores are still not caught, still relatively immune to human claims. Fill them with knowledge and with crowds and they still escape us, outrunning us like the sunlight on the water. Specifically, this age which is able to measure everything but mystery, might tell you just how capacious the oceans are. They comprise two thirds of the earth's surface; they have a close relationship to the atmosphere and are in large measure responsible for our weather; and we know, with the assurance of conquerors, that if all else fails we may be able to save the human race from dying of thirst and starvation by extracting water and food from them, providing our atomic wastes do not prohibit it. We are also learning how to mine the ocean floors for their minerals, how to harness the tides, and how to use their depths for concealment.

Oceanography is one of the great modern sciences and it has revealed mountains, rifts, plains, and canyons on a scale that would astonish us if we saw them on earth, as it has also brought us more knowledge of marine animals at all depths. It has made great contributions to the restless modern mind. How can we look at the sea without at some time thinking of our earth's submerged geology, gigantic, uneroded by wind, sun, or rain, in calm waters inhabited by strange aquatic lives?

Strange is still the word for them. No amount of assessment of the sea's contents quite translates them for us. What, for example, is a fish? What is that flat creature the skate lying there on the sand, with its tough hide and the small slit of a mouth on the same side as its belly?

There is an aquarium at Woods Hole with a collection of many of the kinds of fish that inhabit the waters off Cape Cod. They seem foreign, weird, almost unexampled when you see them in their captured state. I saw a woman standing in front of one of the windows looking at some toadfish, little fat animals with great mouths, squat, with round-edged fleshy fins that gave the appearance of warts and knobs, expertly camouflaged in varied patterns so that they can at once sink in and become a part of the bottom: "Oh!" she cried. "Horrible!"

All the others there become more than the term "fish" when you see them suspended behind glass, floating in their own world of water, strangers in the perfection of their own remoteness. Their dull jaws open and close as they breathe. Their filmy, diaphanous fins wave lightly and loosely. Their flicking eyes pass you by, with a kind of self-enclosed abstractness, a stiffness, as if they had not seen you at all, and no doubt the blurred human form means very little to them. The glass separates the world of water from the world of air. Their bodies curve deliberately and slowly, and then suddenly switch into an unsuspected quickness, while we tourists shove and crowd and gawk from our unbridgeable distance.

At other windows the rays and skates, with fins fused to bodies like wafers, wave through the water. Bottom fish suddenly disappear in puffs of sand. The lean, long sand shark, primitive, tough, swims with infinite smoothness back and forth, an expression of coldness, an incarnate simplicity.

They are all unknown, not of our race, and giving the unknown the old credit of fear, they *are* horrible, monsters in their realm, with intercommunications, receptions, that we are unable to touch.

An aquarium is a luxury. Most of the fish we see are dead, a boatload of wet, cold, slippery white and gray flounder, cod, or haddock just come into port, or dying, like a striped bass caught by a fisherman casting off the beach—flipping on the sands with all its cool brightness still alive, a slippery, lucent sea green. The color loss is quick as a fish dies, leaving the rippling shades of its great medium behind.

The world of ocean color comes inland in the spring with the alewives that migrate from salt water up inlets, streams, and estuaries on both sides of Cape Cod. They are silver, like the sea they come from, with backs of gray green, and in a shallow stream they seem to reflect the colors of the season, having in fact the ability to change the pigment in their skin so as to blend with their surroundings. They mouth the water and stare forward with their big eyes, running upstream with the unswerving directness of their need to reproduce—which gives us at least one reassuring alliance with them!—and being of a fairly large size compared with most fresh-water fish, they have a look of marine capacities, a fast-schooling fish made for water masses, great sweeping currents, and tides.

Even the alewives, which migrate by the hundred thousands, are only suggestive of the far running but hidden nature of the oceanic depths. Most of us, failing a glass-bottom boat or a glass-sided submarine, have to stand on the beach and take in the vast motions of the sea surface with only the vaguest idea of what is happening below. Sometimes it looks like a bowl of dazzling, dashing light, and at others a gray, monotonous range under a raw wind with white-groined waves constantly moving across its distances. The sea takes all the light and air, the storms, clouds, moon, and stars, in endless, various reflections over its watery reaches, with a monumental acceptance.

Are there not a thousand ways to describe the sea which in their sum amount to inscrutability? How can you translate its abundance even by counting so many thousands of protozoa in a drop of water? Who can fathom the range of appetite it contains, the ferocity of the life its amplitude allows?

One day in early fall I traveled from the Cape with a party of people in a chartered boat, heading for an area some ten or fifteen miles out. The offshore breezes coasted over smooth, sun-bright waters that carried some of the land's litter with them, sticks, leaves, petals, and even butterflies. At one point a dragonfly skimmed past us; and silky seeds of milkweed and dandelions went sailing and twisting by to land eventually where they could never take root. Farther out, oceanic birds like jaegers, shearwaters, and phalaropes began to appear. When we were plowing out across the open ocean with its short-crested waves we came upon a broad path of waters which were foaming and flashing and leaping, a white windrow of fish flipping violently above the surface, lasting perhaps a mile or more. Evidently we had come upon an area that was rich in plankton, attracting many small fish, attacked in turn by larger ones. What we were seeing was part of the classic food chain that leads, in terms of size, from microscopic plants and animals to whales. The sea was splitting its sides with riches, and a kind of savagery that most of us hardly dare admit, although as a race we are not so far removed from it ourselves.

As the glass on the aquarium window separates the specta-tor from the world of the fish, so the long nearly unbroken line of the Outer Beach stands between us and the vast, alien reaches of the North Atlantic. It is not *our* natural environment, and so we can legitimately call it treacherous, sullen, cold, and grim, and even in its hours of brilliance and warmth it seems to lead us off in no terms we can call familiar. It is full of fickle changes, fogs, and storms, unpredictable shifts in mood. We are still unable to set forth on the open ocean without the skill of a sailor or the protection that a technical civilization affords us.

Yet our neighbor the sea provides the amplitude and even, being still relatively unaffected by human ownership, the re-generative power of what is both dangerous and undiscovered in the universe. All its shores are washed by a capacity. If it is constant in peril for us, and for its own voracious inhabitants, it is also beneficent as a medium for life. Those tidal rhythms, watery colors, and reflections are translated into living organisms whose uncounted numbers are assured by their vast and rel-atively temperate home.

We only see a small part of those numbers, at least consciously, since sea water may be swarming with invisible life, but during spring, summer, and early fall, the sea's bounty often reveals itself. Countless moon jellies for example, pulse through waters inland of the sea during the springtime or in Cape Cod Bay, where I have seen comb jellies in great profusion during late summer. Watching them, it is not only their primitive, brainless nature, or their numbers, that has seemed incredible to me, but their approximation to their environment.

It has been estimated that jellyfish are 95 per cent water. Dried out, they resolve into almost nothing. How could such evanescent creatures be predators, killing and ingesting living organisms? When you see such transparent flower-animals it is even difficult to believe that they have the nerves and muscles to be able to pulse through the water; but their chemical balance, their physical responses have a direct relationship with the sea water, whose salts are in them. Salt water is a liquid medium for life, a blood that circulates through the creatures of the sea. So close is the association of the sea and its lives, though each species has its unique kind of locomotion, respiration, aggression, its own way of feeding and being food, joining in the employment of energy, that it is almost tempting to inquire whether the sea does not have an organic nature of its own. I will not get very far by suggesting that a medium and environment "knows" anything beyond what all nature knows, but this primal "mother" great provider and provided, has its own deep rights in the realm of being.

In summer and into fall you can see thousands of small fish schooling in the shallow tidal edges of Cape Cod Bay, moving slowly until approached, when those closest to you swing forward, or run, rush, and circle as need be, the whole crowd sometimes escaping with a simultaneous, sideward sweep. They are all spontaneity, life on the run, endowed with limited attributes from the point of a "higher animal" but of strict extravagance in form and action, born of ocean waters. They suggest the incomparable, swimming out of range.

There is something of this suggestion in many specific aspects of animal, or even plant, life in the sea. In a sense their fascination lies in what has not yet been discovered about them, but just as much, from the average human point of view, in the way their actions are those of the sea rather than the land to which we are accustomed. In fact all of us are obliged to make surface discoveries a great deal of the time, even with respect to what is around us, or even inside us, like fishermen following the seasonal movement of fish, sometimes predictable but often hidden and unreliable, or students who chase after migratory birds in planes. So the sight of grunions wiggling in California sands, depositing and fertilizing their eggs, bound to a complex interrelationship of spring tides and the moon, still excites our curiosity, being a phenomenon that is not fully understood, taking place in a proximate but different world.

Migrant fish, like the alewives, may return not only during the same season each year but very close to the same day as a run of the year before. Perhaps the cycles involving sea and climate average out very accurately, but it is too complex a phenomenon to say that it goes like clockwork. Tides are measurable but constantly changing in time and amplitude. Environmental conditions in sea water are various and the sea's coordinate relationship to the atmosphere is an elaborate one. Rhythmic response in an organism may be simple and spontaneous—like a fucus, or rockweed, only ready to spawn after a period of exposure at low tide—and it will have its causes, but the causes themselves are greatly complex in nature.

The sea's discovery will not be made by factory ships that process their huge catches of fish, by killer submarines chasing after whales, or by mining equipment. We can physically affect its life with our one-sided power, but it will remain protean and indifferent and we will go on imagining our conquest of it.

On this overdiscovered and overexploited earth the sea remains a wilderness, a resource not of goods but of what is rich and wild. That which we have been unable to use up, or harry to extinction, has the power to renew. The sea is a positive mystery. I hear the surf's continual breathing in the distance; I see the stars that literally cover the sky over the beach on a winter's night like white animal plankton in the spring waters; and I realize that I know no more about them than I know about myself. The depths are still ahead, with the fear and the temptation that the undiscovered arouses in us.

All of us are drawn to the sea's edge as to a fire. Its vast reaches roll and heave in the light. There is an incalculable weight of waters withheld just beyond us, a roaming kept in check. What an exalting thing it is to see those waters dancing with silver castings from the moon! Even in our careless, civilized state, drinking beer, watching driftwood burn, or absorbing the sun and one another, in no way obligated to the kind of cold suffering or exile which sea and seashore have meant to men in the past, there is something in us that wants this brilliance, this barren waste.

The sheeted surfaces blown over by all winds rove on with their freight of light during the day, constantly changing, sometimes black, purple, and gray under pigeon-silver skies, with hazy, soft horizons, sometimes silver scudding with gold, or blue, green, and white in all shades; and always the tidal balance, the surf's fall and drag at the sand's edge, whatever the season.

During the autumn and winter months the cliffs hang their shadows over the beach very early in the afternoon, cold darkness moving toward an iridescent surf that reflects the last light of the sun. The sunset shows curly salmon and fiery orange streaks on the other side of the vast flat table that often runs with sea ducks at this time of year; and then, singly, the stars begin to shoot up their spears and arrows, alignments for eternal navigation.

IV

A Rhythmic Shore

On the beach it might be said that there is no such thing as decline and decay, although in a physical sense drastic change is obvious, from year to year and even from minute to minute. In a northern forest where the trees have been left to grow for many years, I have sensed the presence of a great establishment, something silent and absolutely personal, a society of trees with its own strong relationship to the sun, to the roaring winter winds and snows, to dry years and wet, using the earth-bound materials of growth, decay, and old age as provisions for indefinite residence. These tree communities culminate in "climax" formations, dominated by particular varieties such as maple and beech, or spruce and fir, to progress no further until some great interference, such as a lumbering operation, or climatic change—an increase or decrease in average temperatures over a period of years—may start a community succession all over again.

On the other hand, the beach and its cliffs that stand as buffers against the sea never allow much in terms of residential time, except to societies that can adapt themselves to living between the wet sand grains, minute plants, and animals; and beach hoppers that burrow in on the upper parts of the beach, or other crustaceans that sink into the sand and out again as the waves go up and back, reacting simultaneously. It is a terribly exacting place to live in. Life is short. Disturbance is always to be expected, and the more so in the course of a storm, which may change the whole physical character of the beach itself.

While I was walking on the beach I rented a small summer cottage in the South Wellfleet area during the late autumn and early winter months, so as to be able to spend nights as well as days by the sea, and I paid it sporadic visits when I could. I remember one night when the sea showed me just how candidly elemental and violent it could be. A northeast storm had been making up all day. Off the Provincetown area, where the waters are protected by Peaked Hill Bar—extending from Race Point to High Head, some thousands of feet offshore and parallel to it—the sea though gray and choppy, was relatively calm, while the wind blew hard. I could see several fishing boats on the horizon. They were surrounded by clouds of gulls. The sky was not totally overcast to begin with but full of handsome blue-gray clouds that sailed across the air like great round slates. Farther south the gray Atlantic foamed and rocked ahead, and the green surf came in dashing with spume and spray, pouring an angry froth on the shore. Finally the sky closed in completely.

By nightfall, water driven by air filled earth and sky. A little ship's bell on the porch outside kept tinkling, and the wind rained blows on the house. The walls thudded as if they were being struck by rocks. Rain pelted the windows and the cold knifed in between the door and the sill. The sea was putting on a profound and concentrated roar. I went out and fought the wind as far as the top of the bank above the beach. Beyond and below that it was almost impossible to stand. A mountainous milky surf was seething, overturning, and piling in. Fury was riding high. The wind belted houses, shrubs, and scanty trees. The beach grasses were tossed, bent down, and released. Rain slashed and whipped wildly everywhere and it seemed that all the natural power and danger in the world had been let loose. When day broke majestic breakers were booming and pounding down the beach as the north wind drove long lines of spray across their heads.

This is the kind of storm, not infrequent between September and May, that flings down ladders reaching to the beach, undermines or tears away the asphalt parking lots, throws wharf pilings and great ocean-drifting timbers around as if they were matchsticks, and leaves them strewn on the sands. It also tears away tremendous amounts of material from the cliffs, as well as straightening or leveling out the contours of the beach. The cliffs are eroded by storm action primarily, not by the tides; but after a series of storms uncovers a part of the beach, displacing great volumes of sand, sections of the cliff may come down by gravity slippage, because they are not supported underneath, and high tides may help the process.

The extent of cliff erosion is very variable, and in so far as storms are concerned, depends on their degree of intensity. Offshore bars and shoals protect the beach from the action of the sea to some extent. When they are breached during storms, the result is a greatly increased cutting away of the beach sands and erosion of the cliffs. When bars reform and build up again the beach slowly recovers its former volume, though what the cliffs lose, of course, they cannot regain.

The estimate given for the average rate of cliff erosion along the Outer Cape is from two to four feet a year. I have heard of one family who have had to move their cottage back three times during the past forty years, a period in which the cliff, so it was estimated, may have receded nearly 200 feet in that area; and their house lot was not extensive enough for any more moves. Most residents or returned visitors can remember some change in the topography of the cliffs over the years. Not long after the end of World War II, when I came to live on Cape Cod, there were still the remnants of the old twin lighthouses above Nauset Light Beach, in the form of a curved brick base at the top of the cliff. As time went by it was undermined, then started to slide down, reached the base of the cliff to be completely buried by sand, but was uncovered again some years afterward. In South Wellfleet water pipes still project over the cliff, indicating the presence of summer cottages some forty or fifty years ago.

Changes in the beach are more immediate, and not likely to be so irretrievable, but even there it is possible to see its fluctuations over the years. There is a great rock off Nauset Light Beach that used to stand high and clear at low tide some years ago, but it has been undercut and filled around with sand and recently only its top was showing.

This is not a level, stable, protected kind of beach. It is steep, full of long shoulders and curves, and fluctuates in outline not only as a result of storms but with each tide and even with every wave, making new bays, curves, shallow hills, and hollows; but the beach is an interbalanced system. All its materials come from offshore or the erosion of the cliffs. Wave action removes the cliff material, and currents moving parallel to the shore take it both north and south: there being a neutral point around Cahoon's Hollow, halfway between Highland Light and South Wellfleet, although its location is dependent on the angle at which the waves come in along the shore. Half the cliff material moves north to build up the hood at Provincetown, and half moves south to be deposited along the sandspits from Nauset to Monomoy.

A study made by the Woods Hole Oceanographic Institution, under the direction of John M. Zeigler, points out that the north and south ends of the Cape terminate in fairly deep water, 205 feet off Race Point and about fifty feet off Monomoy, and that: "It seems unlikely that material is moved to the Outer Cape from deep water, either from north and south, or by lit-

toral drifting from any other part of the New England coast.
Drifted detritus would be trapped or obstructed many times
before it could reach the beaches of the Outer Cape."

During the course of the same study beach profiles were measured for several years and it was found that the sands were constantly changing in elevation, all the way from several tenths
of a foot in one place during a mere ten minutes to a ten-foot
loss in another during a period of two days. The average change
per tide was about four tenths of a foot and sometimes went up
to a foot.

The beach has a kind of rhythmic beat, up and down. If its
changes were translated into visual, continuous motion on a
screen you might see it dipping, rising, and undulating like the
waves at sea. Turbulence and change are not outside a frame of
order. Loss is balanced by gain, so that the sand which is taken
from one part is added to another, and though the relative
volume of the beach is greatly reduced it may be restored in a
year or so to more or less its original size.

Zeigler's report, incidentally, makes the observation that the
beaches "become very steep and full in summer and are quite
variable in winter, spring and fall" characteristics governed by
the "sea state" during those seasons. Sea state, if I understand the
term correctly, refers to the offshore characteristics of the sea
surface, the height, length, and steepness of its waves, and their
velocity, all governed by the wind in its many different phases.
The waves that cut the beach away during fall, winter, and
early spring are characterized by their steepness. On the other
hand the summer waves that build up the beach, although they
may be the same height as cutting waves, are not steep, the long
swells that you see offshore in the warm months being typical
of this kind.

From Nauset Coast Guard Beach to Highland Light the cliffs
range between 60 and 170 feet in height, and they are made of

the stones, boulders, sands, gravels, and clays of what geologists up to now have called an "inter-lobate moraine," meaning the mixed glacial material built up as a ridge along the sides of two moving lobes of ice—in this case two lateral moraines joined as one.

A new study by Dr. John Zeigler, which accompanied his work on beach erosion, puts forth another theory for this area which is that the ridge was already there before the glacier came. It caused the glacier to split into two lobes and the material it left behind was fluvioglacial outwash, there being no real glacial till such as makes up the moraine before Nauset. A carbon dating taken in this upper Cape region puts its age at 20,700 years.

The Lower Cape, from Orleans to the canal, is a true terminal moraine, having material that was pushed ahead of the glacier and left behind when it melted north. It is characterized by uneven hilly country full of rocks and stones merging with a slanting sandy surface on the south which formed the outwash plain.

The cliffs may only be eroded in substantial amounts during storms, but to a slight extent they are always eroding. In some sections, especially during hot and dry weather, there falls a continuous stream of pebbles and granular sand, made a rich reddish-brown by iron compounds, looking in the strong light like a broad rain of precious metals, treasure chests broken open. In other places sheets of fine sand pour down in miniature Niagaras, or flow and fly ahead along the cliffs before the wind, having the look under slanting winter sunlight of light smoke from many fires.

Chunks and fragments of clay are loosened by the weather from their beds in the cliffs and are often washed by heavy rains so that a gray liquid flows and fans out for some feet across the sands. Occasionally boulders will loosen and tumble down. In fact small stones are constantly falling, rolling erratically part way down the beach and leaving their tracks behind them. The cliffs are the prime source of the beach's materials and a repository of the ages that preceded it. They have a proud and vulnerable role in a context where everything is subject to displacement and removal.

Taking an average of three feet a year, the Outer Beach may have required 1760 years to erode a mile in width, even though that is one of those general figures which may mean nothing so far as detailed geological history is concerned. In any case, not only cottages and lighthouses have gone their way but also such topographical features as marshes and ponds, with all the frogs, fish, and plants that belonged to them. On the cliff tops and very close to the edge, there are many glacial kettle holes, now dried up, but once full of water instead of sand, so numerous in some areas as to make one uninterrupted dip and rise after another. On the Nauset Coast Guard Beach, where the cliffs have ended and are replaced by a long sandspit protecting the Nauset marshes behind it, there is good evidence, jutting out on the beach, of a former kettle hole, showing a fine dark sediment composed of organic material which once lay under beds of peat.

The cliffs' glacial material, in whatever form they were left on Cape Cod some 12,000 years ago, was part of the land's erosion, of geology's rising and falling history, for countless years before that. Since then it has been constantly exposed, loosened, easily eroded and ready for the taking, by winds, tides, and waves, but all of it was changing and movable in terms of the great stretches of earth time. Many of its stones and boulders were being wind and waterworn, cracked by frost and heat, long before they were plucked from hills and ledges, transported and left by the glacier to give the Cape its present form. Now they are being broken out and rolled down to be worn again. Like the tides, they are part of a balance, a flow, and containment, that is prodigious in its reach.

The cliffs erode; the surf churns the sand; currents carry away the sand and other cliff debris; storms cause the sea to break in across sandspits and bars, so that they change constantly in shape and position. There is a magnitude of effect involved at this meeting place of sea and land. It is a magnitude that stretches between a sand grain which may be less than a millimeter in diameter to storms whose force makes man-made explosions of nuclear energy minuscule by comparison.

Sand is perhaps the apex and symbol of the whole process in which the existence of the beach is involved. It is moved and shifted grain by grain in the displacement of its masses, lifted by waves, carried by currents, and set down again. Sand in the evolution of the beach is not a static material but an agent of dynamic energy, following out the motion of water and air, itself their product.

Sand grains, which are of great age, have been worn down from rock and the mineral grains that make it up, to particles, largely of quartz, with some feldspar, that are sufficiently durable not to be reduced to the consistency of mud. The wind which moves the waves and is the ultimate cause of all beach movement, also may have a more important effect than water in abrasing and rounding out a sand grain. The action of grain against grain is more abrasive in the air than in water, which acts as a cushion. In any case a sand grain made of quartz reaches a nearly irreducible size after a long period of time. It might eventually be reduced to powder, but it is now protected by the grains next to it because of its small size and the film of water surrounding it. This water, held there by adsorption, is what makes it possible for tiny animals like nematodes and copepods to exist in such an environment.

Pick up a handful of moist sand and it is heavy and relatively cohesive. Through a hand lens you can see the grains fall off in pearly clumps. On the other hand, dry sand is blown down the beach in its separate grains like rice, and sorted on different levels according to its weight and size. Each sparkling grain is an entity unto itself. It is easily lifted and moved by the energy of waves and currents and at the same time heavy enough in the mass to give beaches their malleable stability.

A sand grain is a product of earth, with beauty, quality, and dynamic character, shining clear in eternal process. Sand has the strength and resilience needed to hold up against the violent tonnage of the waves, and at the same time to share in their employment. It is always being remolded into new shapes by the art of wind and sea, shifting restlessly, moving from age to age. What we call the inanimate not only has its weights and measures but also a wonderful proportion with relation to the forces that send it on. It has a going out that is as rhythmic in its way and as full of viable light as the migration of organic lives.

V

Dune Country

Sand dunes, as distinct from sandspits, or the banks at the head of the beach, are found in a few restricted areas on the Cape, but their two primary locations are the Provincetown hook and at Sandy Neck in Barnstable, on the Bay side. Inland of the beach, far enough not to be exposed to the constant wash of the tides or to flooding seas during storms, the dunes have forms and motions of their own. They were originally produced by the wind, and it is the wind that reshapes them, blows over their shoulders and down their slopes, making mounds and ripples on their surfaces, and also undoes them and makes new ones again.

The Provincetown dunes, which I had passed by on my June hike down the beach, represent an exposed region of several miles in extent, uninhabited for the most part except for a few gray beach houses perched on the dunes overlooking the sea. They are continually being added to by sand which the dry northwest wind picks up along the shore and blows inland. Because of its dryness, this wind also has the greatest effect in moving the dunes. Damper winds causing moisture on grains of sand, make them more resistant to being moved.

Much of the region is held down by low vegetation. Its sandy reaches are patched everywhere by Hudsonia, or beach heather, pitch pines kept down almost flat on the ground by wind and salt spray, and its slopes and hummocks kept intact by beach grass; but in other areas, and they are extensive, the dunes have broken loose and roam like the waves of the open ocean, with great crests and long, deep troughs. They look as if they should have a slow, massive momentum of their own, but they are moved by the wind, migrating in a west to east direction at the rate of some ten to fifteen feet a year, creating a considerable problem at the point where they skirt the highway across from the town of Provincetown. On the far end of Pilgrim Lake high dunes loom over the highway and are continually drifting down on to it, hardly deterred by snow fences and the planting of beach grass, so that the sand has to be cleared off frequently.

It is a young country, even compared with the rest of the Cape, which, in geologic terms at least, is by no means an ancient land. It is postglacial and is made of material brought along the shore and added to a reef of glacial debris. It begins where the glacial material of the upper Cape ends, easily seen where the cliff at "High Head" breaks off above Pilgrim Lake, and then it stretches and curves out very close to sea level. Samples of material taken in the area showed a carbon dating of 5000 years, comparatively recent times. Also there seems to be good reason for believing that much of the dune country was broken free and set to wandering by the hand of man.

Between the dunes and Provincetown there are a number of ponds, marshy areas, and woodlands, including some good-sized stands of beech and oak. These woods must have been considerably more extensive at one time. In the dunes that now hang over them there are remnants, tree trunks, and stubs protruding through the sand; and there is at least one part of the dunes that seems to show evidence of a wood fire that took up a big area, though when it occurred is not clear to me.

Thoreau wrote about the dead stubs of submerged forests projecting above the surface of the sand in the "desert," as he

called it, and of numerous little pools in the sand filled with fresh water ". . . all that was left, probably of a pond or swamp." He may have exaggerated these pools as an indication of former ponds or marshland. They are located a little above the water level which extends everywhere under the dunes, and so are likely to be found at the bottom of the dune troughs, or hollows between the dunes. Some of these pools, or fairly long and narrow stretches of shallow water, may stay in much the same place over a long period of time if the levels where they are located are at least partially held down by vegetation. They are filled up by rain water during fall and spring and then dry out during the summer months, but where the dunes migrate before the wind, they also travel behind one dune and before the next; and they are seldom deep enough to develop typical swamp vegetation.

Thoreau tramped the area in 1849, and two hundred years earlier the dune area on the town side of the "Hook" and possibly further must have been much more circumscribed and held back. The early inhabitants cut down all the trees they could find, for firewood; "try works" for melting whale blubber; boats, houses, and salt works (in the days when salt was produced by boiling sea water instead of the later refinement of using solar heat to evaporate it).

Blowing sand became a threat to Provincetown and its harbor early in its history. In his *Cape Cod; People and Their History,* Henry Kittredge describes the war declared by the people of the town against almost every stick, living or dead, that surrounded them.

"When the Mayflower band arrived," he writes, "the sand hills to the north were for the most part held stationary by trees and shrubs. But from the earliest times the inhabitants, following the example of visiting fishermen, fell upon the trees until the sand lay bare, a prey to the four winds of heaven. The captains of fishing schooners were allowed to take sand ballast from these hills, and not content with this, the citizens turned their cattle loose to graze on what clumps of vegetation still struggled for existence on the denuded hills, with the result that the grass was demolished as fast as it grew. The sand was free to blow down upon the unprotected village with every northwester, threatening even to bury the houses.

The danger attracted the attention of the Colonial Government as early as 1714, when an act was passed to preserve the trees. In 1727, Provincetown was incorporated, and a dozen years later another act forbade the pasturing of cattle on the sand hills. The Court might as well have forbidden the winds to blow or the sun to shine. Provincetowners cared nothing for laws, and continued to cut wood and turn cattle loose for the next hundred years; in short, until the danger, instead of threatening, actually arrived. The sand buried a house or two, and was advancing toward the town, salt works, and harbor at the rate of fifty rods a year along a four-and-a-half-mile front. In 1825, another commission was sent to study the situation and suggest remedies. This time they found the citizens so frightened by the marching sand that they were ready at last to obey the laws. They planted beach grass on the barren dunes, kept their cattle in the pound, and stopped cutting down young pine trees. Thus was the sand anchored and the town saved.

Pilgrim Lake is what is left of East Harbor, an extension of the main harbor of Provincetown that ended in marshes separated by a narrow strip of beach on the outer shore. The sea was a constant threat to this barrier and the people of Provincetown were afraid that it would eventually break through and start sending tons of sand into their valuable harbor, eventually making it unusable. A dike, 1400 feet long and seventy-five feet wide was finally completed in 1869, cutting across the mouth of East Harbor at the entrance to Provincetown Harbor, so that both houses and fishing industry were no longer threatened with burial; but the dunes, though held in some control, have continued to blow.

There is a small hill called Mt. Gilboa on one side of the highway at Provincetown, facing another Biblical peak called Mt. Ararat on the other, and if you climb it you can overlook the harbor and the roofs of the town, as well as the dunes and sea in the other direction. (Provincetown, incidentally, consists of a belt of houses narrowly strung along the inner shore with its streets directly oriented toward the harbor, appropriate to a people whose trade and thoughts were toward the sea. This is also true of the houses, which were built longitudinally, parallel to the streets.) In the fall, clam diggers bend down over dark flats at low tide between stretching fingers of water. Dories are stranded in the mud, or move gently on low water. Beyond them are the curving, stockadelike enclosures of the fish weirs, and draggers move in to the mouth of the harbor out of the bay. The sunlight fires the sandy faces of the long, low cliffs that extend down the inner shore of the Cape.

The town, which is so thick and crowded with cars during the summer months, a host to the cities, teeming with talk and color, a variety of human shapes, sizes, and exclamations, so reclaimed that you can hardly conceive of its austere past, becomes diminished again to a mere cluster of houses, a tenuous edge on water and sand. On the far-going Atlantic side, the dunes billow and toss. The Ararats are everywhere, peaks, crowns, domes held down by yellow beach grass on the mounds and hillocks from which the slopes dive down.

As the world's dunes go, these may not be of major size and extent. On the other hand they have been measured at heights

between sixty and eighty feet, and at times dune ridges may
have reached elevations up to 100 feet. Also, their scale is such,
leading from one open face to another, that human figures climb-
ing a steep side across an intervening slope of no great distance
seem tiny. The walls keep looming up and the valleys dip between, so that the whole landscape is full of a wide motion.

In all this bare largesse of sand, the texture is clean and clear.
Shadows move over it like loving hands. The wind's touch in
turn has made grooves, grains, and ribs on the surface. In some
areas the black mineral magnetite joins with garnet to make
blackish-purple ripples in the sand, or irregular masses, or
little brushstroke feathers and clouds. Everything shows clearly,
from human footprints and the long ruts made by beach buggies,
to mice or rabbit tracks. And I suppose that in the summer—if
you pounce in time—you can see insects leaving their traces,
like dune grasshoppers, colored and grained like sand, or a spider
that buries down in the sand, thus avoiding extreme temperatures; or even a toad. I once found a Fowler's toad quite far out
on the beach where it must have wandered away from the dunes.

A stick that drops down from one of those shrubs so besieged
by wind and sand waggles down a dune making a fine tracery,
or what looks like a stamping of birds when it is lodged in one
place and blown back and forth. An oak leaf merely blown for
a slight distance down the sand makes a track, with all its lobe
ends imprinted like a long tassel or thin strands of separate
strings. Except for the beach-buggy tracks, that follow one route
fairly consistently, and may be visible for months at a time, and
the beach grasses, continually renewing their precise circles on
the sand, most of these tracks soon disappear. There is a constant
moving of sand particles, a sweeping over by the wind. The open
dunes are trackless areas where tracks take on great significance.

During winter days when the northwest wind blows with fury
along the exposed shores of the Cape, it may be too uncomfort-

able to stay in the dunes for any length of time. You gasp in the polar air and hide your face from the stinging sand. Each sand grain is lifted and sent with the speed of a projectile along the surface of the dunes. Given a little shelter from which to watch you could see the dunes change shape in an afternoon, or an hour. It is on days like this that they migrate like waves, with long slopes on their upwind sides, steep ones on their lee.

On their bright and stable days, the long dune shoulders at the top of each rise tilt you up, body and vision, into the dizzy heights of a sky graded from cobalt to indigo, the way the scale of things in the landscape goes from sand grains to rocking seas without distraction. The dunes almost seem to ask for a long-distance running from both men and clouds. They are a place of flying, falling, and tumbling, shaping the motion of what comes to them, asking for an approach that soars.

Also, they have their secrets, their ground-level associations. In October the beach-grass heads are loaded with yellow seeds. Where the plants are clumped together, providing protection from the wind, nests of seed gather on the leeward side, visited by birds that leave many little tracks and sometimes a feather or two. Mice also leave their dimpled trails, circling around the beach grass, traveling across bare sand for short distances before they disappear. There is a special delicacy in the visits of birds and mice. I had the fancy, following these small trails, of watching mice under the moon, with all their scuttling, nibbling, and investigating, so that some of their excitement, their fidgety life dance might be translated for me. I even thought it might help bring me down from a world too heavy with size to a neater reality.

Startled by a little crash of twigs and leaves, I saw a rabbit darting up a dune slope. It bobbed to the top and stayed motionless for a few seconds in a bayberry thicket until I followed it to find what might be the meeting place of a whole tribe of rabbits, if I could judge by the amount of tracks and pellets of dung there were, all on the rim of a small bowl held together by the bayberries with a small scrub oak coming up from its base. These semiprotected hollows are quite typical of the dunes. There are also small woods of pitch pines, thickly carpeted with needles, where the tree roots can get some moisture at the bottom of a valley between the dunes. Scrub-sized oak, pine, sometimes bayberry, beach plum, or wild cherry, hold down many hollows, with the help of beach grass on the shoulders around them.

The beach grass has had much deserved honor heaped upon it, in the proportion that it is able to live with the tons of sand that are also heaped upon it. It is perfectly adapted to being covered over by sand since it sends up stems which in turn root themselves, and then grows on, letting the old roots die. As a sand hill builds up, the beach grass is able to maintain itself in this fashion without being buried and to hold down the sand with a network of roots and stalks. It stabilizes such hills until the point where the wind may sweep so constantly around them as to expose them and cut away the sand, leaving the grass in splendid isolation with its outer roots hanging in mid-air. So beach grass and sand have a special collaboration which man does his best to encourage, especially after he has made rescue work necessary.

The sand masses have great weight and volume and are stable in themselves but it is their surfaces that flow and shift with the wind, so that the whole region is remolded over periods of time. It is fascinating to sit in a valley between the dunes and reconstruct their curves, seeing how the sand has been swept down one side and blown up another, sent over a hill to make a new one on the other side, held for a long time and then broken loose to change its residence, motion, and stability joining to make those noble forms.

The dunes may threaten man's house, or road, or wood lot in immediate terms, but in themselves they are like distant monuments dedicated to natural force, perfect, calm, threatening or joining all that which lies ahead of them with equanimity. Time and its lapses seem immaterial, more so than the wind that shifts them. Now, or in years to come, a migrating dune will kill off a tree or a shrub and what does it matter? Can I care about what happens to one of a thousand scrub pines? I think not; but perhaps I can care about the event in the whole sequence of growth, change, and reshaping. Slow and statuesque, the dunes under the great air are another balance in process, like the beach beyond them.

I think of some of the trees in the dunes and their struggle with the winds and the encroaching sands, and I am unable to shed tears over something that is unable to cry, but sometimes the word desperation comes to me, when I see evidence of their long efforts to hold on. You will see a dying cherry tree that has sent shoot after shoot, trunk after trunk, all over the side of a dune or sand hill that is being worn away, and they are full of the contortions of struggle—arrested, like the statue of Laocoön and his sons wrestling with the snakes, but real enough. Or another hummock or small dune, where a beach plum or bayberry may not have enough purchase left, has a mass of twisted branches and twigs strewn down its sides, the wreckage of a genuine defeat.

On the north edge of Provincetown the migrant dunes skirt the woods and thickets on their borders like icebergs, clean-rounded, immense shoulders of satiny sand slipping by trees: shad, bayberry, beach plum, red maple, oak, or pine. Because of the stable nature of sand, except when it is blown, they stay where they are, great suspended masses, their progress only measured at intervals, leaving evidence of trees that are buried, or about to be buried, behind them.

I am indebted to Dr. Loren C. Petry for pointing out to me that some trees are able to grow in the same way as beach grass, while they are being covered with sand. Pines will die when they are only partially buried, but this is not true, for example, of cottonwoods whose branches send down roots soon after they are buried, and so maintain their water and mineral supply. He has seen fifty-foot specimens of this tree—along the southeast side of Lake Michigan—of which some forty feet were buried, with the remaining ten feet growing vigorously.

The trees in the wooded areas bordering the dunes, particularly the pines, look as if they were covered with a soft whitish powder. It is caused by the very fine sand grains dusted over their leaves and needles by the wind, and during the winter this can be seen for miles down the Cape, well south of High Head.

Almost all the trees here have a temporary existence, holding on as well as they can, fighting for light, food, and moisture. Even if there used to be more woodland than there is now—and the evidence is good—there is nothing about this narrow area, stretching into the sea, made by the sea in collaboration with the wind, that looks settled. The word stabilized can be applied to a dune and in a sense to anything that remains rooted, anchored, or in place for a certain length of time, but in this case the word balance might be better. Motion, either latent or in view, is in equilibrium throughout this rare place, half desert, occasionally wooded, full of gardenlike patches of low growth standing out in their variety of color and shade, seeming to move like the clouds. There are shadows everywhere, made by low twigs, needles, or grasses, the slightest thing lying across the sand, in sketchy rhythmic patterns tossed by the wind, while the greater shadows made by the high dune outlines are shifting steadily with the time of day.

Aside from mice, rabbits, skunks, toads, insects, and the indigenous plants, this seems a place for nomads, and the birds that are free to forage, like a dark pigeon hawk that swoops across on its hunt for prey, or an occasional marsh hawk, breast feathers gleaming in the sunlight, its shadow passing across a dune wall. Little flocks of birds burst here and there through the thickets, like chickadees, myrtle warblers, or juncos that move around on the ground pecking for seed. In their fall migration many of the juncos, or "snowbirds," reach Cape Cod by a long, over-the-water route, and flocks that arrive on the Outer Beach begin to move up into the dunes in a search for fresh water, perennial pilgrims.

I sat on the top of a high dune one afternoon and watched a beach buggy swaying and swinging up in my direction along a track that led from the shore. It droned up and careened by me, plowing and slipping through the sands, and away down a long slope it went on the dunes' free forms, cutting across the shadows that were spearheading toward the sea. Then I heard children's voices in the distance coming over quite clear and shrill, falling off at intervals before the wind. The slopes and valleys stretched with pure travel in between. It was the kind of place where all views and associations keep on, across a shifting range. It lacked fixed ways, decided roads. Only packed in by the open ocean and the long reaches of time, the roving dunes made a continually majestic statement which no amount of cans, broken glass, or human footprints could erase.

Off on the end, the edge, past the cities and the suburbs, the fixed house lots, the fields, and plains that make a patchwork of an entire nation, here is a country let go, barren, down to an essential minimum, but tossing and flowing with its own momentum in an envious proximity to the sea. It is the first and last land in America.

VI

A Change in History

The history of Cape Cod is fairly well known. I say fairly well because I do not see how it is possible to recapture the deep complexities of what was present and now is past, although there is enough past left in us to provide great confusion about the times we have to face. Many tourists run after "charm" or what is "quaint," terms which are slight enough to admit that they have very little to do with the dark realities of three centuries. Now we come and go in great bounds, from great distances. Motion and change make our constancies. We are in no need of staying put. We are attracted by the starlight in the heavens we have created for ourselves. We look on the earth's great flowing beauties with an inclined eye. For all its "conquest of nature," perhaps because of it, our civilization has a tenuous hold on the waters and lands it occupies. We are in danger of being overlords, not obligated to what we rule.

We do not "visit" in the old sense of the word, stopping in for fish chowder, or rum or a cup of tea, nor are we customarily invited in because we are tired and out of our way. There is no time for that, and besides there are too many of us.

The new human plantings do not fit the old outlines. Cape Cod is now subject to a population spreading out as a result of the tremendous growth of cities and towns. It is predicted that the number of winter residents will increase by forty or fifty thousand in the next twenty years, and the summer visitors to the Great Beach may pass all bounds eventually. As the speed of transition has been increased between one era and another so has our individual speed, in arriving and departing. When you buy a piece of land on the Cape you do it as an investment, as a kind of fluid security, not for its own sake or something too priceless to let go. There are always other places to move to. Each man used to be his own nomad, now nomadism is supplied to all of us by the mechanics and riches of society. During the tourist season the average length of visit per person has been estimated at three days, enough time to sense the breadth of things if not the circumstances.

If we are all to be itinerants, wasting and leaving, or suburbanites, Cape Cod will have a hard time keeping what open beauties it still displays, even with the National Park, which has saved a great deal of it from the seemingly unalterable army of bulldozers in the nick of time.

The record, written all over the Cape in the form of cut-over woodland and wasted topsoil, does not say much for human foresight at any time, with or without the bulldozers. In that respect we have not changed, though we are not as dependent on the locality we live in as we used to be. Food and resources come from afar. Still, all places, regardless of the human adventure, have their underlying tides, their own measured and perhaps measureless pace, and they shade their inhabitants in subtle ways. We continue to be affected by what we can neither transform nor avoid. No amount of dry ice stops the hurricane. We have no barriers to keep off the arctic air. So those of us who live here still complain helplessly about each other or the weather, while ghosts of penury and puritanism still haunt the local houses.

The area in which I stayed for that brief nomadic period of my own, was filled with cottages, on slopes ending on the cliff above the beach, a majority unoccupied but with a house

here and there showing a little more substance to it, the evidence
of a year-round resident. With some exceptions, they were bare
in appearance and devoid of individuality. No uncommon effort
had been made to give them much distinction. In the winter
and fall they lost whatever color by human association they
might have had during the summer. Some of them were flat-
roofed, pastel-painted little boxes without even the virtue of
exposed wood, and since they were not in Florida they could
not borrow any youth from the sunshine. Their spirit was old
before they were built, and in that respect indigenous to the
seashore. The bare coast and the gray waters seemed to hold
them in contempt, or at least indifference, and they became
as gray themselves. They are due credit for their lack of preten-
sion, whether planned or not. They did not take up the landscape with improvements and cultivation. They sat on their
own little plots of sandy ground, with a few pitch pines, Hudsonia and scrub oak, joining the general economy of the landscape, no blowing leaves and limbs above them, no spreading
lawns around. Whoever might live in them after the mild,
money-making season could be gripped by the real weather without interference.

Our age may give the lie to all those who are interested in
antiques, even if there are any old ones left. Perhaps there is
no alternative if we have to get to the moon or bust. Will there
ever be such a thing as an antique rocket? But there is still a
flow of age, a distant sense of things that it is possible to find,
hanging like mist over an inlet, booming like the sea over the
far side of a hill.

You can still walk the Old King's Highway in some areas,
a single-track road where it is easy to imagine a horse and wagon
or a stage, during the years when it took two days to get to
Boston and the sea route was the preferred one. Even with
the jet planes droning overhead and the cars grinding gears in the

distance and the about-to-break sound of the future in the sky somewhere ahead, it is as ancient and distinct as the outline of an oak tree. Just its narrowness is enough. I spent half one afternoon trying to find it in one part of its extent, and at last there it was, quite clearly, just the right size for the eighteenth century, with narrow ruts in sandy ground, lowered, indented, washed out in some places, grown over in others, but a ghost with definition.

In the Wellfleet and Truro areas you can still see how the houses were located here and there along the old highway, or dotted around in sheltered hollows back of the beach. In the wintertime you are very likely to meet no one, since there are comparatively few year-round residents. Once the place was full of local need, local talk, or tragedy. What wrecks now occur along the treacherous offshore bars can usually be taken care of by men of the Coast Guard who can get to the area quickly in a jeep and sound the alarm by phone. When there was no radar for ships, hardly any means for wide and quick communication with authorities on land, localities were responsible for the wrecks that might occur off their own shore. There were volunteer lifeboat crews composed of men from neighboring houses, with a boat kept ready in a hollow above the beach, ready to be launched out to the rescue, in terrible seas that were a common part of existence.

In the early part of the nineteenth century Cape Cod towns had between three and four hundred sailing ships between them and a majority of their men went out to sea. In a great storm occurring in October of 1841 the town of Truro lost fifty-seven men, being already burdened with a large population of widows, and on the day after the storm nearly a hundred bodies were recovered along the Cape Cod shores. Most of them were caught while they were fishing on George's Banks or were making a desperate trial of returning home, with a northeast gale screaming and the sea sweeping their decks.

The bars off the Outer Beach from Peaked Hill to Monomoy have been responsible for an incredible number of shipwrecks in the past, and taking the measure of the storms that strike the coast, it is hard to see how there could have been as many survivors as there were, even with the gallantry and local experience of the amateur lifesavers. Many ships ran aground too far offshore to be reached, and were pounded to pieces. The death-dealing power of the offshore sea in these storms seems unparalleled. The surf has the turmoil and roar of an avalanche. It chews and churns at the cliffs taking great volumes of material away so that it seethes with foam and sand, the masses of teeming waters plunging in, heaving and conflicting, an amalgam of unapproachable violence.

Many of the lights that welcomed sailors, or warned them off, are now gone from the headlands and from houses along the shore that no longer have to worry about their men any more than they have to worry about themselves. The mackerel fleets are no longer thick on the horizon. The wharves are gone that used to take in the mackerel at Wellfleet on the Bay side. No one eats salt mackerel any more that I know of. I have a friend who spent his boyhood in New York State who was given salt mackerel to eat on Sunday mornings. It had been soaked in milk overnight, having been taken out of a "kentle," which was a small wooden keg, the top wider than the base, about a quarter of a barrel in size. His observation was that it was much too salty a dish for his taste.

The talkers at the livery stable, the central store, or the barbershop are also gone, as well as the sea captains who retired at the age of forty-five or fifty to become big men in their communities. The horses, truck gardens, fish heads, rum and rum runners are gone too, and what old men still whittle boats for the tourists on the beach? The ancient marvels who used to gather Cape Cod moss on their backs, telling hilarious stories about chicken stealing, cow "dressing" (manure), boundary disputes, occasional romantic murders, and hard days at sea no longer seem to be available for reference. What a lot of solid objects seem to have gone from the world!

Perhaps I have left history behind too soon, saying, in effect: "Choose what age you like. You may find yourself in another." Perhaps it is no fault of mine.

During my autumn and winter walks I did find a lasting pleasure in recognizing old things, reconstructing neighborliness, even from a distance, learning to see the silence—the growth and shape of things, the riches of "slow time." The ponds especially, in the Wellfleet and South Truro regions, protected by the woods around them and the land leading up to the cliffs above the beach, were clear and deep and seemed to reflect quiet habitation over a long time. The water lapped on sandy shores in the sweet, airy winter stillness, broken by the loud, bright braying of blue jays. Coon tracks were sharply etched on the shallow margins where they had gone fishing for freshwater mussels that left meandering traces on the pond bottom. On the far ends in the shadows there were occasional ducks, like blue-winged teal, mallards, or scaups.

At Gull Pond in Wellfleet one January day there were scarfs of ice along the shore, and out in the center herring gulls flew up and settled down on open water where a light cold wind broke across the surface. Wavelets were continually pushing and jostling broken ice so that it made a high singing, almost bell-like sound.

Around these ponds were crows, evidence of owls, wintergreen leaves to taste, and wind whisking through the pines, or oaks still carrying dead leaves. I heard the odd little hornlike note of a nuthatch as it was rounding the scaly plated trunk of a pitch pine. Pale light moved through the woods and across the hollows. Silvery trees bordered gentle mossy roads, their tracks loaded with fallen leaves. It was all in a special Cape proportion, colored silver and gray, like the Atlantic, or the herring gulls, the clouds and the sky, or an old house that suddenly showed up in true style and balance, not to be imitated by any century but its own.

Then I walked out to see the great green breakers roaming in, and to hear their thunderous bone and gut fall across the length of the beach. The sound held and it took away, a monumental assurance of power past all the roughness and directness of the old life, its quiet suspension in the present, and the wrenching of the not-yet born.

What you have to face after all, in this low wooded land, in the continual dip and rise of its contours, is consummate change, the way the beach itself, or the dunes are changing, keeping a general state for a minute, or even a lifetime, but quite beyond catching. Its history is water.

Water created it in the first place. When the last enormous glacier melted back leaving its indiscriminate load of rubble out in the sea, it had also created a profusion of holes, basins, gullies, the "kettles" which are now dry or semidry hollows, bogs, or still holding water as ponds and lakes, and valleys, broad and narrow runs with outlets to the sea. At one time Cape Cod must have been streaming with water like a whale's back when it rises to the surface. Now many of the original streams, rivers, and ponds are wholly or in part dried out, but without too much imagination you can fill the landscape with water all over again. Scientific exactitude, geologic reconstruction, make it possible to confirm your sense of the place as full of remnant and abiding fluidity. There is hardly a piece of land on the entire peninsula that does not suggest this.

It is water thousands of years behind, water inseparable from the motions of the future, a power roaring in and destroying, pushing, grinding, ebbing back. It is water in the rain; water in the deep, still ponds; water in the underground darkness; in the gentle seaward running streams; in the tidal estuaries and marshes lowering or flooding over; as sleet; or snow; in icy gales full of the howling emptiness of the winter sea, when the cold metal of the wind pounds on your back and cuts at your face, as it sweeps down the semifrozen sands of the beach where the green and white surf fumes in, rolling and churning with impersonal passion.

Even now the history of Cape Cod is a history of enduring weather, of the same exposures. Only our terms are not the same. Some years ago I stood on the high hills of North Truro late one afternoon, watching the sun's red path shining and moving across the wide waters of the bay, thinking of sea surfaces moving over the round earth to its poles, and the poverty of the winter world around me, stripped to ultimates, everywhere exposed, and exposed to everything. The round hills were so bare that the little separate houses in the distance, down in hollows or perched on the long slopes, seemed to shiver. They glittered like so many frost flakes in the air. I had just come from Provincetown and seen a dragger unloading its fish, and the fishermen cutting them up with red, raw-meat hands. The wind was shipping up the water. The gulls were crying over the racing, lathered shore.

It came to me that what had brought me here had not so much to do with a feeling for the old Cape, with its churches in their simple New England grace, or clam-digging, beach-combing, old wrecks, driftwood, or fish weirs, real as it was in me, but a great new outwardness, a universal human event. Each man undergoes a series of changes during his lifetime in a sequence of experience that corresponds to that of the world. He has in him the revolutions, the escapes from holocaust, the interspaces of peace, the fact of war, the anxieties, the cry that his being be fulfilled, the never-ending human examination and measuring of things. So I found myself to be "way out," a Cape Cod term anteceding the Beat Generation, and meaning far from your home base, with very few old promises behind to sustain me. I had to come to terms with an age without age, a locality without location, perhaps a divinity in fires of no precedent or name. Above all I was required to change, to face in new directions.

The gulls floated in the cold air with customary ease. On my way home I saw a great blue heron flying over a marsh and inlet, its broad wings spread out like a cloak, long legs stretched straight behind it, with feet curled up stiffly, head and neck crooked back. Then it landed in shallow water. Its wings folded and it stood straight up, with a surprising, statuesque height and gaze, the long neck and head above a flock of ducks that were swimming and feeding near by, assuming the kind of composure special to a race of herons that would serve indefinitely. The wind ruffled the water, swept over reeds and curving grasses, sending the last light of day roving in splendid colors over the entire marsh.

All the measured lights and shadows of day and night, the tides of the sea and the tides of the season, the response and joint association of all life's components in that place stayed much the same as they had ever been, in spite of the way we hurled in our roads and relocated ourselves without rest. Its natural order was still there for old expectation to seize upon; though in terms of accumulated knowledge and wants it was more complex than it had ever been, and would have to endure a human association that was itself on the waters of change, holding hard to the mechanics of its coming. Cape Cod had suddenly lost a slow, accumulative history, perhaps in a matter of twenty years, and would be treated like the rest of the world—as it happened, as it would come about under human auspices. Our problem, one of many, might be this: how could we reconcile universal commitment with the inviolable nature of a single place?

VII

Barren Grounds

The oceanic landscape reaches across the round earth, over a curved horizon, and that may be one reason why men keep returning to it. The sea attracts the experience of distance. There is still some vicarious adventure to standing on a cliff, breathing the far-ranging air and imagining ships hidden by mists on the horizon, or unknown lands beyond that, or even remembering lands once visited. Over there is where the great passages of history have gone by.

As recently as fifty or sixty years ago, man and sea were involved in a more personal alliance on Cape Cod, and its seamen once voyaged around the world. At the same time there were some local inhabitants who considered it a major expedition to go from one side of the Cape to the other. The fishing, shipbuilding, and voyages to foreign lands that was more characteristic of the Cape before the Civil War than after it gave what might have been a too narrow community, concentrated only on its own affairs, a healthy connection with the rest of the world.

Since the Second World War Cape Cod has been filled with relative outsiders, many of whom have been transported, not necessarily through any fault or wish of their own, to stations around the globe. A place that once went out for its sustenance now waits for the world to come to it.

One of the few people I met during my off-season walks on the beach turned out to be a man who had retired from the city. The open air may have been conducive to revelation, because he told me a great deal about his life during the ten or fifteen minutes I talked to him. It turned out that the place where we stood had some significance in his own history. He looked out to sea from the edge of the cliff and pointed out over the water to show me the general region where transports used to gather during the First World War on their way overseas. He had been on a Navy escort vessel.

"This country," said he, "is waste," as he talked about war, small business, rough competition, lumbering, and all the size and circumstances of the men and societies he had met and fought and endured. Through a life-long experience of waste—or waste space—and all his tired compliance with authority and anger against it, he had saved room in him for voyages. He told me that he had come to live near the sea so that he could walk along the cliffs and the beach whenever he wanted to, and to look out, I guess, when he wanted to with a relatively free command view of destiny.

After I left him I met another reminder of war, spread out for several miles along the tops of the cliffs. It is now within the boundaries of the National Seashore Park, and one day, when the beach grass takes hold of its denuded areas, it will no longer be recognizable as a military reservation, but when I first walked through it Camp Wellfleet had just been formally disbanded. Although it was completely deserted, its buildings and some of its installations were still intact. It had been an antiaircraft post, and not of primary importance to a coast which was not likely to be attacked, but I have heard local residents speak of the constant, annoying sound of practice firing, which made the walls tremble and the dishes fall off the shelves, and for several years after the war ended fishermen use to protest that their boats were in the line of fire.

The camp was in what geologists call the Wellfleet Plain. It was on these bare levels above the beach that Guglielmo Marconi built his wireless station and sent out the first transatlantic message in January of 1903. The year before, he had built an elaborate structure with twenty masts, and this had blown down in a heavy onshore wind. The successful message, which took the form of an exchange between Theodore Roosevelt and King Edward VII of Great Britain, was sent from only four masts, which had more stability in Cape Cod weather. It is typical of the Outer Beach that although Marconi transmitted waves that crossed the world, the sea has had the last word. On the day I walked through nearly sixty years later there was nothing left of what he had constructed but a few fallen bricks on the face of the cliff.

Marconi's towers were long gone, but the Camp Wellfleet lookout towers and firing range were still more or less intact, and the place only lacked occupation to make it come alive again. The public had been kept out of the area for many years, but now I could walk in on a winter's afternoon and not meet a soul. I passed a sign saying: MILITARY RESERVATION NO TRESPASSING, not without vague qualms, and memories of my own months in an Army camp, half-expecting the sound of "Halt!" to ring out.

"Yessir. Yessir." I said to myself, starting to prepare my excuses to some ghost of past authority.

There was no sound but the surf and a pelting rain, that fell on bare gravelly ground seared everywhere with tire tracks. Bareness was something the Army brought to all its posts, so that a bunch of grass was considered unnecessary, or tended for dear life. The Army city, once a humming, purposeful anonymity, was now completely silent and alone, but for me it still kept some of the power of its restrictions, arousing old apprehen-sions—that tightening of the stomach at facing some new un-known. The bare white barracks were still intact, and the power lines. There were signs indicating underground cables, or la-trines. There were off-limits signs on empty streets.

I stood in the rain and remembered that essential order, with its own enormous kind of waste and consumption, and the feelings of frustration and boredom it produced in me. I re-membered the routine, the rote-mindedness which often passed for efficiency, the utter helplessness that many soldiers felt dur-ing wartime, and were obliged to accept, about being part of something huge, anonymous, even reckless and uncalculated, an ignorance of which they themselves were ignorant and to which they had not been invited. I also remembered the un-assuming friendships you could make in the Army, the directness with which men accepted each other.

A sparrow hawk flew over. I noticed deer tracks on the ground. They were interruptions of a nature that did not con-cern me very much as a draftee in an Army camp, although—more than most—we were exposed to the wide nights and their stars, the wonderful freshness of dawn, and the extremes of heat and cold. There is a naked timelessness to Army life that allies it to a sea. A soldier's life was restricted and oversimplified —he was not his own agent—and at the same time he acted for the world, cast out on an open plain. A great waste took him, equal in its surface or its depths, in being out of his hands. When he protested, he was protesting against the passage of all the nights on all the waters.

I can remember a fellow barracksmate one evening after dark saying he had something of great importance he had to speak to me about. We went out and talked in the company street, standing on the sandy grounds between the buildings, conscious of a towering night with flashing stars. He talked desperately, on and on, about the life he had been planning before the Army took him away; he complained that he and the girl he was to marry had been put off; he talked bitterly about the job which had now been denied him, the business he was going to establish, and: "Why? Why? Why?" What business was it of the President of the United States to start a war and send him into it?

It is murderous not to be able to fight back. It is also appropriate for the Army to denude the ground of its grass, the beach grass that holds it down above the cliffs. It is appropriate for the sea to roam on with a blind eye, and for the cliffs to fall and the sands to shift and blow. It is inevitable, at one time or another, that each of us should stand on these barren grounds. The gloom of the sea puts all other darkness and gloom in jeopardy. Its brilliance is impenetrable. It carries light over the earth's surface like a turning crystal. It is overbearing and restless and at the same time as strict and balanced as its tides. Perhaps it is best approached in misery of soul, because then it stands out in all its cryptic mastery as the raw room that owns us, the desert without illusion.

Camp Wellfleet had eight towers, spaced along the top of the cliff for several miles. Watchers could look out from their transversing positions over the coastline and the sea and signal the accuracy of the antiaircraft gunners who fired at mobile targets over the water. I climbed two of the towers that still had ladders. They were in fair condition, but clearly not too long for this world of wind and spray, of ice, rain, and snow, and the fierce summer sun. Most of the windows were broken, the wires ripped off the control boards, and the floors, with boards splintered or gone entirely, were littered with wire and broken glass. A cold wet wind whined through. I wondered how many young men had felt cast off, lonely, and bored on this lookout over the dark sea. Some of those on duty had left their names behind, probably after the war was over, judging by the dates: Sweeny, Morton, Yarborough, and they also left the names, portraits, or disfigurements of their girls, or would-be girls, the signs of need in wastes of order.

Concrete gun emplacements and bunkers were still intact, with empty cartridges and ammunition boxes on the ground outside. A strand of barbed wire made a little clanging sound of unused warning as I brushed by it. Toward the far end of the reservation, on the Eastham side, I passed another off-limits sign and sat down on a ring of sandbags located in a little hollow on the very edge of the cliff; they were beginning to slide down the face of it like Marconi's bricks. Looking down on the beach where blackbacks and herring gulls were the only sentinels, facing in to the wind, I thought of how many worlds, how many inventions, how much devising we had run through, at a faster rate even than the sea cut down the cliffs. The maniacal weight of one war had gone, but the knowledge and power it let loose had sent us on, committing us to our human ends in the most inclusive and at the same time isolated sense, universally vulnerable.

The wind sent dark clouds of ruffled waters along the sea surfaces, surfaces that tilted and flew, stretching away and disappearing, and the sky light, feather gray in the rain, reflected everywhere. The long surf line sounded with the crash and rattle of stones. The vast flow went on unhindered, restless and controlled, delivering and holding back, a nay and yea sayer at the same time, passing all experiments, accepting all possibility without a care. How could the sea do anything about reassuring mankind as to whether or not we would survive our own acts and commitments? Did man make war, or did war make him? Perhaps we love the sea for its denial of us.

Sitting on the sandbag, I thought of the GI who had last been there, manning a gun now replaced by missiles and rockets—

bothered perhaps by the cold, penetrating wind, feeling useless, waiting for his discharge from the Army, wishing he were somewhere else, not knowing beach grass from seaside goldenrod, or one gull from another, but knowing the sea, with its one sound.

VIII

A Landscape in Motion

There are a number of elevations on the Cape from which it is possible to see both sides, getting above intervening houses, trees, or hills. On the same Wellfleet Plain where Camp Wellfleet was located the moraine tilts all the way down from the cliff above the Outer Beach to the shores of the bay, and reaches of land and water come into view from all directions. One plane leads to another by easy transitions. The cliff tops shine in the wind above the steady pouring sound of the waves and the dancing of molten gold and silver on the sea. Beach grasses glitter. The land ahead is full of coarse scrub oak and green patches of bayberry moving toward dark green woods of pitch pine and clusters of houses, reaching the sheltered shores of the bay beyond them, with salt marshes, gold and red; water-shining, brown tidal flats, and a rim of blue water on the distant horizon.

It is a stunted land, not overhospitable to life by the looks of it, although flocks of chickadees bounce gaily through the scrub as if giving it their free acknowledgment. As the autumn progresses the reds change to brown, plants darken or die down, shrubs lose their leaves, and the grasses bleach. In all seasons it is a place of low growth, ready in its hardy way to receive what the wind and sun can send it. The sky is very wide overhead. You can see from one tidal area to another—almost from one climate to another—standing on the bare ground. In scale the view approximates what you can see from high in the air.

A plane shows you a much wider panorama, while diminishing the land, eliminating the size of locality and local things. It takes you high enough to see the curve of the earth, the concrete highways like ribbons across the country, the thin lines of roads and streets, the checkered fields, patches of lakes, and sprawling cities. A jet plane cuts across time. You can run after the sun as it falls on the other side of the world and almost catch it, following the mountain shadows over America, and since you pass time in that sense, not able to go faster than the speed of light, but crossing the rhythmic stations of earth and sun, I have felt it as a longer journey than that involved in a car or train. What might ordinarily take days is reduced to hours, but when we landed I have felt the days in me as much as the hours. We bypass the clock. We go from low to high, bridging a gap between the individual and the universe, leaving earth's confinements for indefinite space, but local time is still inside us.

On the ground, obviously enough, you limit the horizon by the extent of your vision, and the horizon in turn limits you, but land and water are held by their relationships to space and to each other. Apparently all climatic cycles are world-wide; and the immediate, local weather is in part dependent on the weather behind and ahead of it. In the same way the only limit to the landscape is the globe itself. Its reaches go out of sight, if not of universal measure.

This seaside country often gives you the feeling that the sky is the limit. One opening beyond the trees, another mile revealed, and the earth and sea from the top of a dune, the world you stand on, may become exalted in its scope. Perhaps people climb hills and mountains not only to get to the top, or as an activity in its own right—reasons often given in answer to questions that may be of no great value—but to join the range of the world, to be up and outward bound, and above all to have a sense of the unities in and beyond them. A greater landscape means a new communion.

I once climbed a small mountain in Maine with a group of Sea Scouts. We stopped just below its summit, where there was a bowl surrounded by rocky heights and slopes and holding clear, cold water, the size of a small pond. The boys stripped and went in swimming, and all their excited yelling as they jumped in and out of the water resolved along the rock faces and deep crevices into echoes that rang and choired—heard from above—like *Te Deums* in a cathedral. And far down and around for hundreds of miles were the houseless mountains flaming with color.

One of the boys asked: "How many acres do you think there are?"

For all its matter-of-factness, his question brought us in touch with massive distance, an over-all light and wind above the great carpets of color, a landscape running with power, having a latent silence, a prodigious weight and matter.

Mountains or seashore make for revelation. So on this sandy, tilting peninsula sight can keep on going. On one side the head-on majesty of cliffs, beach, and open sea, and on the other, calm low headlands facing sheltered waters, two different environments, with the west wind blowing over and the clouds flaring and shifting in the sky. You are in the lap of the waters, the balance of the tides, and in the arms of the weather.

Each patch of ground, varying in the degree to which it is receptive to organic life, is a complexity of substance and influence. The weather that circulates over it, and in terms of light, relative moisture, and varying temperatures invests it too, has its seasonal constancies but it is always in a state of change. Cape Cod feels much of the time as if it were two-thirds wind, and people with touchy nerves might well think they were being pushed by it in directions they were unable to go.

The Cape has a maritime climate, somewhat milder than the

mainland. There is no use exaggerating its mildness since it can feel as cold or colder than the rest of New England when the northwest wind takes its uninterrupted course through the ribs of the land and sears its way along the shore, but, in general, annual temperatures are slightly higher. In central and western Massachusetts, in New Hampshire, Vermont, and New York State, the average number of days between the first severe, killing frost in the autumn and the last one in the spring has been estimated at 180–210. For Cape Cod, on the other hand, this is 120–150, the same that prevails in a thin coastal belt south of the Cape to Virginia and North Carolina where it widens and starts west across Tennessee.

The waters to the south, in Buzzards Bay and Nantucket Sound, have a higher annual temperature than the waters of the open Atlantic along the Outer Beach and in Cape Cod Bay, a southern extension of the Gulf of Maine. On the other hand the waters north of Cape Cod, though cooler during the summer, tend to be warmer during the winter, because of the depths of the Gulf of Maine and their heat-carrying capacity. Cape Cod Bay, and Buzzards Bay have more sea ice than any equal area on the coast of the United States with the exception of Alaska. Sustained cold during January and February often results in weeks of pack ice stretching off into the Bay as far as the eye can see, at least from the level of the shore. This extra touch of the Arctic off the Cape is due mainly to a combination of cold winter winds from the continent and shallow water.

The difference in average water temperatures between one side of the Cape and the other may have its effects on the local weather. During the fall especially, when cold air moves over the waters of Nantucket Sound they may be covered with fog, whereas it can be bright and clear over the Bay, only a few miles distant. The normal kind of fog occurs when warm, moisture-laden air moves over cool or cold water, and is quite common in spring and summer. When a cold, dry air mass, on the other hand, moves over warmer waters it may result in what is called "Arctic sea smoke" a kind of wispy, steamy fog in turbulent, rolling air, rising to ten or fifteen feet above the surface of the water.

During the winter Cape Cod is also subject to rapid changes in temperature depending on whether the wind comes from the northwest, with cold, dry, continental air, or from east and south off the ocean, the latter being seldom below the freezing point.

The tip at Provincetown has much the same temperature as the sea island of Nantucket. On the other hand the town of Barnstable on the lower Cape may have an average summer temperature which is sightly warmer than Provincetown and a colder temperature in winter, since it is that many miles closer to the interior. I have driven down the coast from Boston several times during snowstorms when an area as close to the Cape as Plymouth was completely covered with snow; and as I drove south the storms turned to heavy flakes of wet snow on the near side of the Cape Cod canal and then to rain as I went on.

The sea's capacity to store up solar energy means that it exercises a moderating influence on the Cape, which is warmer during the winter than the mainland and cooler during the summer. Also, there are less thunderstorms on Cape Cod during the summer months than on the mainland, and the annual rainfall is likely to be lower because there is less showery precipitation, although local residents might be justified in thinking that water was on them much of the year in one form or another, as fog, salt spray, rain, or humidity.

The late fall and winter is often characterized by cold, raw windy days, with the temperature just above freezing or at the freezing point, and the air is loaded with moisture from the sea and sometimes smells of it. During heavy storms the wind drives the salt spray inland with great force, depositing coats of salt on houses, telegraph poles, and wires.

During the winter the Cape seems at times to be caught and tossed between the weather of the sea and that of the continent, but in general the principal air masses during fall and winter come from inland and in summer from the southwest. Winds from the north and west usually bring in continental polar air, which is dry and cold, though it may also arise in part from pacific maritime air. The source regions for many of the storms of early spring and early fall are the Gulf of Mexico and the Caribbean. Most of the severe spring storms, sometimes coming after a fairly mild winter, are the so-called "coastwise southeasters" which blow up the coast from off the Carolinas rather than from the west. They can result in blizzards because their coastal, maritime air if drawn into a low from the continent is cold enough to make snow.

Such simple generalities and fact sampling is not to suggest, like the Chamber of Commerce, that more people ought to come to Cape Cod, but that it is a land like all others, which is influenced by the forces beyond it. It is no more gripped, pulled, and let go by the weather than most other areas. In fact its temperature made it a good place for the first English settlers to find. Think of the Middle West in July, or January, for extremes! Yet Cape Cod has a special place in the wind, an outside hold on the roaming of the seas and the advent of the air.

The tides that rise and fall along this ocean-going spit of land are just as varied in their way as the weather, but more predictable. They accentuate the difference between one part of the Cape and another, and they are responsible for some of its physical characteristics. Great tidal ranges on the north side expose wide salt flats at low tide and allow the development of broad areas of salt marsh in sheltered embayments, whereas along the shores of Nantucket and Vineyard sounds, where tide ranges are much smaller, the marshes and more exposed flats are less extensive.

In Cape Cod Bay and eastward to the coast of Maine the average tide rises and falls about nine feet, but in Nantucket and Vineyard sounds the range is up to four feet at the most, being as little as two feet off Woods Hole and in some of the salt ponds. The time of high water varies also. It occurs four hours later on the north side of the Cape than at Buzzards Bay.

The Outer Beach is an area of transition so far as the tides are concerned, and their range drops steadily from nine feet at Race Point to four feet at the end of Monomoy. These diverse tides, all along the shores of the Cape, are a product of its very shape, and of the coast from which it juts out, astride the submerged continental shelf, whose shallow water also affects them.

It is the nature of waves—and a tide is a wave of a special kind—to move more slowly in crossing shallow water, rising at the same time to a greater height. Waves expend the energy of their motion when they increase in height, an effect which can be observed as they heap up before breaking as surf on the beach. So the tidal wave moves in from far offshore starting with relatively low ranges, some two or three feet at Sable Island off Nova Scotia, with similar readings in Bermuda and the Bahamas; but when it reaches the outer coast of the Cape it is augmented. To the southwest of the Cape the increase is only moderate, the figure for the entrance to Buzzards Bay being three and one half feet; but moving north it gets much higher. To reach the shores of the great embayment of the Gulf of Maine, formed where the coastline drops away north and east of the Cape, the ocean's tidal wave must first cross the shallow waters of George's Banks, a passage that requires more than three hours (which explains the later time of high water in the Bay). In the process the tidal height increases to the nine-foot figure, a reading which is true of Provincetown, Plymouth, and on up to the coast of Maine.

So the Cape lies between two tidal systems, created and separated by its geography. On the south side, incidentally, there is a complex pattern of tidal movement caused by the fact that both systems meet. Tidal waves enter the sounds between the Cape and the islands of Martha's Vineyard and Nantucket from two directions and pass each other. The combined effect of this "interference" results in rapid changes in the time and height of the tide between Monomoy and Woods Hole. Off Nobska Point one tidal wave movement is high, while the other passing it is low. Their interference results in the smallest range of tide (one and a half feet) to be found along the south shore. A similar minimum tidal range occurs off the southeast corner of Nantucket.

I am neither a trained scientist nor an accomplished sailor. I am inclined to use facts for unfactual ends and do not have enough knowledge of the wind not to be tipped over at any time, but if you feel complexity and admire mathematics while in a state of comparative ignorance then perhaps you have some claims on knowing. Most of us have had a hand in observing the weather or gauging the levels of the tide. Weather guessing or complaining is second nature, and on the beach, or by means of the pilings on the wharf, you can guess the tidal range quite easily or judge whether the tides are in or out. On some level below that we have air and tides in us that know the energies of earth from past acquaintance, but we are much too ready to mistrust these depths and to let other authorities do our work for us. Perhaps our natural senses are becoming atrophied. In any case, we do not seem to be sure whether it is the energy of the head or of the heart that we should use for our purposes. But put yourself in the middle of the weather and within the reach of the tides and they sometimes begin to roam in concert in as many ways and to the incalculable extent that you have responses stemming from your brain. All the distant swelling and swinging, the synchronization and intermoving of the waters, becomes as real and immediate as the repositioning of the sun and the changing of its shadows. The over-all wind; the light that shines on the beach grass, moves over the pebbled ground, and sparkles the sea, or turns it into a blazing white cauldron; the knowledge of cold massive depths in one place, warm shallows in another, come into feeling as both unified and infinitely complex. I may fail at mathematics but be an unconscious mathematician, judging galaxies by the ways of light before my eyes.

At my feet, as I sit on the sandy ground on the cliff top, there is a hole made by a spider, neatly defined at the top by a little rim of grasses. Rabbit dung lies here and there. There are a broken puffball, dried leaves, and seeds; and the wind has blown so constantly over the level and open parts of the ground as to take away loose sand and leave a surface of pebbles, which are more or less stable, while mounds and hillocks are held together by shrubs and grasses. These are evidence of a poor community, holding down as best it can, though it is open to migrants and migration all the same.

What lies underfoot changes in a few hundred yards toward vegetation which is a little more protected, and less exposed to violent light and dessicating wind, with low oaks and pitch pines, wood floors, with a certain amount of decaying litter, graduating upward in the quantity of organic life, but the open, exposed, diminished look of this environment also suggests its inherent mobility with all the other component parts of this running world, taking original light and shadow from the vast sky.

The crow with its ragged wings banking away over the tree-tops, the rabbit hopping into a thicket, the fish that school unseen in the salt waters, the man who watches, are all manifestations of a complexity of association and alliance that stops on no single shore. Like our restrictions with respect to the horizon,

we only see, we only live, a fraction of the possibilities allowed
in so great a range; and being restricted, we oversimplify, cutting
life and land down to size . . . a poverty that makes for poverty.

I hear the steady pouring sound of the depths behind me and
I see and feel them rising and falling, taking their inexorable
passage around the Cape. The wind whistles through and like
the in and out of breath lifts and subsides. Field crickets trill
monotonously and faintly in competition with the wind. Crows
call. Seeds blow along the bare ground. A winged seed flies by,
next year's fruition if it lands, this year's providing, perhaps
destined to skim out over the surface of the sea. A flock of snow
buntings swings back and forth, twittering high in the air. Gulls
circle in the distance above a garbage dump hidden by the trees.

In this landscape, here and out of sight, is a mutuality of response, through the sea with its thousands of miles of variety
constantly in motion, and the land besieged by the sea, with
dry and infertile soil, but in a web of tides and climatic influence that keeps its character actively in tune. Like the buntings, or a flock of sanderlings spinning, sun reflecting, diving
through the heights above the shore, the opportunity of grace and
power is always waiting for its use, and nothing that lives and
participates can be called insignificant, from the cricket to the
crow. Diversity is the rule, and each form is exceptional in its
employment.

Through any part of the earth there is a placement, the appropriate condition for plants, animals, the soil, and its constituents, to maintain themselves. The optimum is that there shall
be full use within any given range of opportunity. The more
diversified a living community is the more healthy it is, not only
in numbers, but in complex relationships. Even a "poor" seaside
environment proves this by the very demands it makes for survival. The plants that adapt themselves to it do so by means
both various and precise. Even sand grains have a relationship

to each other in the rhythmic order of wind and waves. The life that comes to these shores, winging in, trying to take hold, blown out, taking semipermanent residence, has its own affinity for place, an organic knowledge of its own part in the physical world. It belongs to an innumerable company with exacting tasks.

Each life proves the need of all others. In a miraculous way, as each natural form is miraculous, the single is also manifold. The rabbit, as it nibbles grass, calls in the hawk. The spider is related, in its reproduction and survival, to the insect it eats. The soil requires microbes to break it down. The growth of plants is directed toward capturing the energy of the sun. Life calls life in the context of earth, water, and sky.

Throughout the wide landscape are a succession of environments, with communities adapting to constant change, characterized by so much mutual attraction and repulsion, so many delicate balances, such a variety of response to influence inside and out that there is hardly a stopping point for attention. We study particular environments so as to predict and understand the behavior of animals, the reaction in plants to variations in the intensity of light, or to relative moisture, or to the chemical constituents of the soil. Each place has its character, its complexity, and bounds.

But environment is more a characteristic of range than a separation in its own right. All migration says so. The division between a pond and its surrounding woodland is fairly distinct. A pond is an entity unto itself. So is the division between salt water and fresh. But the frog that lays its eggs in a pond may travel through the woods during the summer. The salmon, the alewife, and the shad reproduce in fresh water and grow up in the sea. Eels do the opposite.

In a sense each area has its representative, like the water birds, from petrels that spend most of their lives over the open ocean, to fresh-water ducks dabbling among the reeds. There are herons adapted to spear fishing in the shallows; terns that dive for fish in surface waters; others that swim after them under the water. Some of the adaptations are so precise that if the particular food supply of a species is endangered, so is existence of the bird itself.

On the other hand the very distinctness of each species, sharp-billed, webfooted, with gliders' or divers' wings, seems to impart range to countless others, those which exist and have existed, those which may develop in a vast and unknown future. The difference, the space, between a gannet and a dovekie, a great blue heron and a frigate bird, proves all the depths of opportunity.

As I look out on the waters to east and west, to north and south, I either see or envisage banks of fog far offshore, warm summer squalls, biting cold air, torrents of brilliance in the sky, leaping and ponderous deliberation in the waves. Warm air meets me from the Bahamas, cold air from the Arctic, and the migrants pass me as they travel in between. This earth, regardless of man's construction of it, is always re-relating its contexts, playing out new themes ahead.

In this distance, near to far, there is force, and its limits, a counterbalancing as well as intermingling in the land, weather, and tides, and in almost hidden terms the concurrent response of countless inhabitants: the seed makers and dispensers, the hole diggers, the fliers, scuttlers, and divers, those that swim, crawl, or walk. They take part in range after range of consumption and growth, of trials and failures, with endless patience, sudden quickness, flows of energy, going through death and the travel-round of reproduction. They are dancers in a realm that knows where all its leadings are.

There are dynamic secrets underfoot. Lives dawn of which we are entirely unaware. Can we bring ourselves down to their great participation, waiting through dawns, attending the sun, hiding under the reality of wind and storm, where obedience means praise? Here is that universal guarantee of novelty and increase which we try so narrowly to imitate, substituting our simplicity for its complexity, our distressed communality for its balanced crowds, our greed and invention for its terrible provenance. Lord have mercy on us!

IX

Who Owns the Beach?

In the "off" and empty season, after the tides had erased all signs of a hundred thousand human feet, it was hard to believe that the beach could be owned or claimed by any one. It took on the air's cold or warmth, receiving, passing things on, from one day and seasonal mood to another, not as on the land with its plant and animal reactions and obstructions, the hiding; shadowing; coming forth intermittently; but in bold and naked sight, reducing weather to its single qualities.

One day the Cape would be sunny and comparatively warm, and on the next in would come the authentic northern wind, the polar air, roaring and sweeping around with fierce abandon, riotously hard and cold, freezing the ground, cutting at a man, diving on him with an icy weight. The winter wind is so definite when it comes, overwhelming a fairly moderate climate, where roses often bloom late into the fall and hollies grow, as to make you think of icebergs, sliding down from the north unexpectedly to stand hundreds of feet overhead. The sky, threatening snow, writhes and purls up with gray clouds spreading fanwise like auroras, and in the evening the sun goes down with a coppery band on the horizon overhung by a bank of steely-blue clouds as menacing as a shark.

And the great beach received what came to it, retaining its primal right to a deeper breath and regularity, a harsh "poverty-

stricken" environment where man has no lease worth the paper.
It did seem utterly deserted, although the herring gulls and blackbacks flew up steeply over the wind-buffeted waves, then banked
and glided away, and draggers occasionally moved parallel to the
beach bucking the choppy seas, their lines out astern. The
wind threw stinging clouds of sand ahead of it. Except for the
fishermen and the gulls, it was an abandoned world, glistening
wide and cold, lost to importance and sense so far as human society was concerned. For man there is no force quite so inclusive
as his own.

Since the beach is comparatively empty and isolated during
fall and winter, the sight of life on its sands may seem as rare
as a rider approaching you across the desert. I remember what
an extraordinary thing it seemed one afternoon to see a tiny red
crab moving very slowly along, high-legged over the bare slopes
of the beach. I identified it later as a species of spider crab. Green
crabs, rock crabs, calico crabs, and others are common along the
protected shores of the Cape, but out on this stretch of beach
they are rarities. This baby, with its beak, antennae, and eyes
backed and covered by a knobbed and spiky shell, seemed like
an exotic from another world, which in fact it was, having been
flung in by the surf from rocks and seaweed forests in the waters
beyond it. It not only added to the beach, but to me, since it
made me realize that these sands were only shelving off into
further dimensions. The beach is a repository of freight, wreckage, and lives from foreign lands.

This also happens occasionally on land. We all know that the
sea is out there, that the wind swirls over us, and the storms
carry more traffic than planes, but strangers sometimes appear as
if to prove that no place is what it seems to be. One spring a
vermillion flycatcher suddenly appeared in the neighborhood. I
saw it in its exciting tropical gaiety as it flew down next to a
shining patch of spring rain on an asphalt road. It is a native of

Texas and New Mexico. Black or turkey buzzards ride the great airs of spring and sometimes fly northward, wheeling unexpectedly overhead. In November of 1962 I saw a black stork, *Ciconia nigra*, which had somehow managed to make it all the way across the Atlantic Ocean, perhaps managing to stop for rests in such areas as Greenland and Newfoundland. It landed near the Coast Guard Station, now National Seashore Park headquarters, at Eastham, in an exhausted state, to be picked up by the Audubon Society and later transported to the warmer climate of Florida.

The black stork breeds from Central Europe to Korea and China, and it winters in Africa after a long round of migratory journeys. Its advent was greeted with a certain amount of mild curiosity and even some jokes in the local paper, one of which had to do with its liking for Cape Cod scallops on its arrival. What better reason for coming here! (The truth is that like other newly captured birds, it had to be force-fed.) In any case it was a rare event, joining Cape Cod with Africa, and to see it was equivalent to seeing an antelope on Route 6. With large strong wings, attenuated red legs, a long, stout pinkish bill, red around the eyes, it waited in captivity with what seemed to be an air of great sadness, transplanted as it was, taken in to a gray, cold land without any sound but engines, human voices, and the wind, without any greenery but the thin-needled pines; and it roosted silently, twitching occasionally in its inactive unused state, an unwilling, unwitting Marco Polo in New England.

This is a narrow place, restricted by nature and by men, but foreign lives still fly to it like sparks in the air, and the sea beyond it takes things on their way with more room than analogy is yet aware of. What the sea sends in, like a dead skate, a starfish, horse mussel, or finger sponge, seems perfectly familiar as fish, marine, background animals, but they are also genuine primitives, remote not only from human physiology and complete understanding but from that part of the earth's surface that we inhabit. In fact many of the hints of marine life that are either brought up along the beach, or that appear in offshore waters, like a whale or a dolphin, have a theatricality, an off-stage hint of a wealth of other acts, tricks, and forms still to be seen. The simple, primal watery element has embodiments of use which are comprehensible and have been studied for a long time, but these are endowed with physical natures and capabilities that might make an air-breathing, earth-bound human quite envious.

During a violent coastal storm, with winds up to seventy and eighty miles an hour, an exhausted harbor porpoise was cast up on a bay beach recently, and there it died. I confess I had never seen one out of water or even close to me before. For all the pictures I had seen, and all I had read, nothing prepared me for such perfection. Its round body, four to five feet long, was butt-ended at its head, in which there were small eyes, and small teeth in the jaws. It had just as much of the quality of flow as a raindrop, and at the same time was a solid packing of energy. Its skin graded down from the jet black of its back and upper sides through streaks of gray like rain along the sea down to a white belly, and without scales, it had a thick, smooth satiny polish like ebony or horn, perhaps reminiscent of synthetic rubber or plastic but of an organic texture which neither of those products could equal. The porpoise had a single fin on its back and a tail that could strike vertically for power and thrust. Its body was fairly heavy, weighing about a hundred pounds, but everything of speed and liquidity and dashing, leaping strength was reflected there. It lay on the upper part of the beach, conspicuous among the long piles of storm litter, the logs, pieces of broken dories, and thick seaweed, spectacular in its simplicity, a black and white that made me think of breaking waves in the night sea. I saw it curve over the surfaces of the water with consummate grace, slide away, and disappear.

"Where did you ever see more of nothing?" I was once asked as I looked out over endless dry Texas plains billowing like waves. Nothing or everything. Who knows? Who knows what the emptiness leads to or contains? The beach lies open. Its sands and rattling stones lead back through ages of weathering and change and are at the same time part of the wide give and take of the present.

The tiny spider crab, though isolated on the beach, was also a link with a teeming offshore existence, which hid in shadowy worlds of kelp and rockweed, or floated and roamed by with a free energy that was in complete denial of our tightening fall and winter world. Backed by a cliff, walking on sands shadowed and cold, faced by the churning waves, it is hard to believe in a life so rich. There are no rocky shores revealed at low tide and streaming with weed to prove the temperate fertility of the sea. The beach is a transition zone between one environment and another, but except in those areas where the cliffs are reduced to low sand hills, protecting a marsh or estuary behind them, the transition is a sharp one, the sands dipping from the inconstant sky to the constancy of salt water.

Along those stretches of beach where the sea has taken stones and boulders and deposited them offshore, storms sometimes bring in fairly large quantities of seaweed, which need beds of stone for their attachment. The fucus or rockweed, the laminaria or kelp, and some of the "red" algae like Irish moss which are among the more common kinds found along the beach, have no roots, since the plants take all their nourishment from the sea water that surrounds them, but are anchored by holdfasts, stubby structures which in the laminaria may look like the exposed, above-ground roots of some tropical trees, and in the fucus a round expansion of the tissues at its base, which is strongly and tightly sealed to the surface of rocks and stones.

Everything about these weeds, with divided, narrow, or tapering fronds to resist being torn by the waves, with bladders serving as floats, with gelatinous surfaces, with hollow stems, are eloquent of the nature of salt water, its ebb and flow, its depths, its capacious circulation. The seaweeds found on the beach, black, thin, dried out, or fresh and slippery, olive green, brown, or red, having been torn loose by a storm, start growing beyond the violent action of the surf, and grow for the most part to a depth of some forty or fifty feet. Different varieties like different depths, but since they are not free floating unless torn loose they are not found beyond the point where rays of sunlight, necessary for manufacturing food, cannot reach them.

Over and beyond them, in surface waters where the light penetrates before being absorbed, is a vegetation, varying in abundance according to place and season, but of incredible numbers over all, the one-celled microscopic organisms that are the basic food of all the seas. The seaweeds are simple and primitive in structure compared with much of the plant life on land, the more hazardous, contrary environment, and the members of the phytoplankton (the planktonic plants), even more so, although the diatoms, which form a large part of it, show a variety of outer form. Each diatom has a skeleton, made largely of silica, an outer shell hard enough to resist easy dissolution when the plant dies. It is formed like a pillbox, or a casket, or it is shaped like a quill, a ribbon, or rod, or it is joined with others in beads and chains. Each is minute, an etched, crystalline perfection, and each is lost in other billions, which we might only see on occasion as a green or greenish-brown stain across the water.

The shells of dead diatoms rain down through the water and form thick deposits on the floor of the sea. The cliffs above the beach are full of them. Cities have been built on their fossilized shells. In their number the diatoms balance the magnitude of the sea. In size they are basic to the existence of the minuscule animals of the zooplankton that feed upon them, and are eaten by larger animals in turn. A diatom's delicacy and sparkling beauty as it reflects the light could indicate that universal productivity must start with a jewel, and perhaps end with it too.

That which is minute, like the diatoms, or cells, which are the basic structure of life, is a clue to the significance of things, leading from the simple to the complex and multifarious, but finally rounding us back to where we started. A man himself is the unique single cell with its own nature. Each life has its irreducible quality. I have been told that if you look at a diatom through an electronic microscope, from one increased magnification to another, you can see all its protuberances and layers disappear, and finally a sparkling crystalline form is revealed, like a cosmic surprise.

I suppose it is part of my fate as a large and clumsy animal of the mammalian order, crashing through the underbrush, knocking down trees, and displacing earth's other inhabitants, to miss a great deal, at least with my unassisted eyes. To learn about some new form of life which I may have been passing by for years is often something of a redemption. I can then say that we have not yet been so run down by our own traffic that we have lost the capacity to see.

Not long ago a colony of bryozoans was pointed out to me, at least the gelatinous crusts of the compartments in which they lived, like little tufts and fringes attached to the fronds of seaweed cast up on the beach. They are tiny colonial animals that make cups and compartments joined together in branching stems, from which they send out little crowns of delicate, filamentous tentacles waving in the water. There are three thousand marine species of them, growing in different forms, and having different surfaces for their attachment. I had thought previously that the little pale-colored, branched tufts were a part of the sea-weed. Now another small marvel had appeared on my horizon.

The beach was empty where I walked, except for bird tracks, tidal wrack, driftwood, bits of shell, or a finger sponge in evidence of the life alongside it, and depending on the warmth and receptivity to life that the season held, excepting also whatever microscopic animals might be crawling over wet surfaces around the sand grains. Again, emptiness, or poverty, is always qualified. After all the copepods, the nematodes or thread worms, and other groups unseen or unknown to me might be underfoot in vast numbers; and as I continued on there was no counting the number of little holes in the sand made by beach fleas or sand hoppers. As the autumn deepened I supposed they were unoccupied and deserted, since these beach dwellers, as I had heard it, should have been tucked away in their burrows by this time, with the door shut above their heads, waiting for March and April to bring a warm sun which could tease them out of dormancy. But one bright morning in the middle of November I saw a great many of them hard at work.

At first I noticed thousands of little mounds on the surface of the sand in a strip some six to fifteen feet wide along the upper part of the beach, following in general the outlines of the previous high tide. Where a log or shelving bank was in the way, these mounds, and the many holes accompanying them, about knitting-needle size, were concentrated on the seaward side. I noticed that shore birds had attempted to pluck the occupants from their holes and had reached down two to three inches. I scooped out the sand where a hole was, spread it around, and revealed a little animal not over a half inch long, with two large eyes covering the sides of its narrow head. The eyes were not only conspicuous, they were also startlingly white; and the sand-hopper's body flattened on both sides, was a mother-of-pearl, somewhat translucent. This odd creature, one of a family in the order of amphipods, is called *Talorchestia megalopthalma*, a title that gives special credit to its eyes.

I put my pale-moon animal back in its hole, but to be held and thrust against its own volition apparently immobilized it, so I let it go free down the sands. After a second or two it made a few big and seemingly crazy hops—on sidelong springs like a toy— down a line of mounds and holes, popped into a hole and promptly disappeared.

I noticed that little spouts and bursts of sand were coming from many of these holes and with a little patience I could see some of the hoppers coming up as if to look around, as is customary with gophers and chipmunks, and then turning around and going back down again. What they were doing of course was a major job of digging, passing the sand up from one pair of legs to another and throwing it out the hole with a jerk. There was hardly time or inclination to pause and look around the far horizon. It was work that had to be done unceasingly, between tides and between seasons. Perhaps, if tomorrow brought consistently freezing temperatures, they might not appear again in any great numbers until spring; but their usual daily round meant frenzied feeding at low tide and after dark when no winged predators were around, followed by another return to the upper beach and another furiously energetic period of digging homes for themselves. Terrestrial animals, which might drown after a period of immersion, and yet bound on this strip of sand to the tides, they had a more legitimate claim to the beach than most of us.

Looking down at them, or in on their busyness, I had an extraordinary Gulliverlike feeling of encroaching on a world to which I did not belong. It was one kind of an eye looking at another without any sense of whether it was seen in turn or not, in a dichotomy of function, race, size, and place. It took the beach out of my possession. This was a place of other-world connections at which I could hardly guess. Do we need to wait for the men from Mars?

These are extravagant animals, with their grandiose if relatively blind eyes, with their feats of digging, their hunger dance. In a sense they have a very narrow range, between upper and lower tide, between one season and the next, between feeding and digging on their strip of sand, between hiding and emerging, and their life span is short; but what a use they make of it!

Talorchestia megalopthalma is now on my life list, as the "birders" put it, a pearly prodigy of moon leaps that may, for all I know, be the beach's foremost citizen.

I also caught a glimpse of another little animal as I turned over a piece of driftwood. It had numerous legs (seven pairs in all, I have learned), and a flattened body, though slightly rounded on top, and oval in shape, reminding me of a pill bug or sow bug, one of my most familiar landed neighbors, which can be found under almost any boulder or log that provides shade and moisture. The marine, or beached member of the family I met, was grayish white in color, and apparently had the same preference for moisture—if not too much, since it evidently lived at the high-tide line, and was "terrestrial" like the sand hoppers. Some of these isopods swim in the open sea, others live in shallow water, or at the low-tide line, and most are scavengers, feeding on dead animal matter.

All these and countless others are symptomatic of a tidal range, an ebb and flow that extends between sea and land in terms of millions of years of emergence and adaptation. In them the two worlds find their division and also their meeting and intercommunication. Their characteristic areas, their "life zones," from the tropics to the poles, all require extremes of risk and of the struggle to survive it. In one place or another they dance to the inexorable measure of things, limited in what they do but exceptional in their way of doing it.

On this beach, so unique, so well defined, and at the same time so widely involved, every upward surge of the waves and every bubbling retreat sinking through the sand, every range of tide, from the new moon to the old, every storm, every change in the season, every day and every night, is embodied in existence.

I would think it presumptuous of me to claim any more on behalf of a bug or myself than we could in our honest natures fulfill, but faced by the shining tides of life, I am sure we have great things to do.

My translations are on this beach. I am still a part of its measure, and when I forget those overwhelming controls that human power insists on, and all the artificiality men use to overcome their natural limitations, I begin to partake of this miraculous context. It is a cold beach, a bitter sea. Covered with cold, the sands impersonally receive the shadows moving over them tall and wide, gradually shifting and easing over slopes and shoulders toward the surf with its continual lunge, its pull and push, displacing the pale light that stands over the beach and gives it a hard winter brightness. The waves pour and foam and bubble up the beach and recede with a rainlike glistening and seething that sinks in, leaving dark stains behind. The middle part of the beach shows long thin lines like scars where the last tides came, part of the never ending drawing and erasing on this tablet of the sea's art. It is all clean, and naked, defined, and at the same time rhythmically boundless, providing everything that comes to it with an inexhaustible dimension. It needs another language, and at the same time no language could really encompass it. In this bold breath and silence moving up, scene shifting, always starting again, there are decisions of sun and waves, of wind and light, that leave me with a true silence, a great room to fill, though it is in my blood and veins, the roots of me to feel, and any companion whom I meet must be in an ancient earth sense completely new, with a freshness made of a million years.

X

Deer Week

The wind buffeted the sea surfaces so that they were loaded with whitecaps. A black and white fishing boat was bucking up and down offshore. It was a bold and empty day. Aside from the two men that I could see in the boat, the shore was a world unoccupied, bright, wide, and cold, one about which the mass of us might care or know very little.

On the other side, where marshes and inlets entered from the bay, black ducks cast themselves up into the wind, and mergansers rode the choppy waters. The bay also ran hard with whitecaps; and the wind with a bare fury roared head on at empty summer houses facing the north, and drove across headlands glistening with bearberry where pitch pines on slopes in its lee would suddenly take the hard air with a swish, rocking and shaking, then subside to shake again. The wind brought the whole north with it and the gulls that hung there or rose steeply into it, were allied with its violence in a way that was hard to understand.

Halfway between these two realms there was a great deal of human preoccupation in evidence. It was deer week, early in December, and the pitch-pine woods resounded to the firing of guns like the hard slamming of doors, and down the highway at least every other car was loaded with hunters dressed in red, and on nearly every sandy side road several cars were parked. Later on, I even saw a man standing on the cliff looking out to sea, and I wondered if a deer might have escaped him in that direction.

Regulations now required that men wear yellow-orange luminescent patches on their backs, so when they all trooped out of their cars like spectators at a football game, they seemed as covered with neon lighting as a city street. In fact many of them do come from cities to the north and south of the Cape, which can now be reached in much faster time than used to be the case, and they follow the same pattern as many of the summer tourists, in and out, fire and run. For those who live away from streets and highways, deer week can seem perilous. The lookouts stand blocking the side roads and sometimes park their cars across them. They troop whooping and hollering through the woods where I live. The guns resound from all points of the compass.

Earlier in the season is the allotted time for shooting game birds. One afternoon I met a number of men who were returning from a hunting expedition on the shore. It had been fruitless. One man had managed to shoot a partridge on the way, but he ruefully admitted that someone had stolen it from the back of his pickup truck. Crowds of hunters started straggling back, while guns were still going off in what seemed a completely indiscriminate and probably frustrated fashion.

"Pretty hot around here today!" said one old man with great cheerfulness.

I was helping one of the hunters extricate his station wagon from a muddy hole, and by that time I had a feeling that, like many other human enterprises, hunting was a communal affair which might turn out one way or another, but like a battle, had no certain outcome. It was clear, in any case, that very few of these men had much of an idea about the habits of the animals they were hunting. Some species of ducks, for example, feed more readily after sundown and so are more easily found, and more vulnerable. A half century ago, the population of wild fowl was probably less safe than it is now. A yellowlegs, flying up out of a marsh in late autumn, did not have much of a chance to start south. Some local hunter was waiting in antici- pation, someone who probably knew the marshes and the shore as his ancestors had known them.

If the hunters had an unlimited season on this narrow pen- insula, Cape Cod would be in a state of siege the year around, regardless of what happened to the ducks, partridge, quail, or deer. We have the universal problem of room and numbers. After all, the human population is increasing at a faster rate than most birds. Perhaps our populatedness results in less concern for the rest of life simply through lack of association with it. Do we know what we are shooting at? Hunters who blast away into flocks of eiders or Canada geese, leaving many of them wounded, unable to retrieve the rest because they are too far out in the water, are not doing anything but getting rid of their feelings, which are not necessarily worth cherishing.

The deer population may not decline because of hunting. Their numbers, their balance between starvation and survival depends largely on the kind of country they live in, on its vege- tation. Cape Cod is only a half mile in width in some parts of it, seven or eight in others, but down the middle of it there is a wide belt of low growth, of tangles, shrubs, and low, cut-over woodland which provides good forage for deer and good conceal- ment, even with the human armies in their midst.

Hunting deer is thought of as an American heritage, our birth- right, part of the Thanksgiving celebration, handed down from fathers to sons. Since deer are one of those species, unlike their predators the wolf and the mountain lion, that have managed to live abundantly in the presence of man, so much so that they sometimes require "weeding" to save them from starvation, hunting them is as legitimate as it ever was, provided the hunting is controlled; but we no longer need them as we did.

Having left the age behind when venison was our essential meat, we now have an odd relationship with the white-tailed deer. In some states more deer are killed by cars than by hunters. They are directly influenced by human civilization. In turn, civilization is dependent on them to the extent that they provide the basis for a multimillion-dollar industry. We think that it is our hunter's right that deer should exist, but we are not the hunters that we used to be. What is a deer for? Guns, gasoline, clothing, ammunition, whiskey?

The fact that they are still wild in the midst of us may be more to our advantage than any claims we make on them. They are afraid of man and keep their distant beauty from him. The heritage *they* keep is wildness, which still has the power to arouse fear in us, and sometimes pity, as we may pity all life, including our own, that is cut short or broken by the inexorable laws of the universe.

On that December day during deer week, full of cold air and the sounding guns, I saw a doe walking across the road, some distance ahead of me and not many yards behind the beach. Two cars had just roared by with hunters in them, before she made her appearance. She seemed either wounded or exhausted, going very slowly, pulling her hindquarters stiffly behind her. When she saw me, that white flag of a tail flew up and she went off the road up a slope into the woods, but with only moderate speed. And then the doe shivered somewhere on the cliffs under the all-mastering winter air, a legitimate prey of men, who turned up their car heaters and sped away.

Later on I found deer tracks on the cliff tops where I walked, and a hollow where a deer had rested and bent down the grass. I could see the hunters sitting or standing all along the shore road, waiting with rifles ready, walking into the woods behind, getting in and out of their cars; and their "ho!," "hah!," or "garr!," sounded across the way. After a while a number of them began to hurry ahead, almost tumbling as they ran, to converge on a deer which had apparently run to the bottom of a hollow. They surrounded the hollow on all sides, many men standing on their car tops with rifles pointing down. Whether there was actually a deer in view, whether it was shot, or managed to escape, I never learned. There were too many guns in the neighborhood for comfort.

The doe moved on slowly through the stunted trees above the sea, not too long for this world perhaps, and the fishing boat—a very rough trade on that day for common flesh and blood—rocked forward through the waves. After a while the darkness began to fall, with a thin smoky yellow and pink band on the western horizon and a new blanket of gray clouds mounting overhead, so that all of us began to turn in under the cold breath of night.

I wonder, in that light which changes for us every hour, every minute of the day, through the wild wastes of the sky, through the countless years of earthly inheritance and change, how we became so overmastering in our numbers and needs, so divorced from the exactions of nature? Shall we meet up only with ourselves?

Perhaps all hunters, those who know their deer, their mountains, and their forests, with an ancient admiration, and even those who abuse a hunter's "right," knowing nothing but confusion, are trying to keep in contact with a natural mortality which our world denies. Perhaps we need help from other animals besides the human one.

Everything in this landscape, from gulls and ducks to driftwood, marsh grasses, and deer, had a vital distinction. The beach with its perpetual reshaping and scouring worked on each stone and lifted each grain of sand, so long as there was stone and sand. The gulls hung overhead, colors fitting the shore and sky.

Even the boat had a fittingness, a sea size of its own, and so with feathers, logs, or purple stones, all in solitary nobility, but swept and washed into a mutual keeping by the air and the tidal presence of the sea. I asked it to show us light and life which was our undiscovered own to help us through our mutual violence and upheavals, our narrow days.

XI

Impermanence Takes Its Stand

Just as the sand bars offshore change shape continually, and the beach loses and gains in volume and elevation, so the plants and trees work so hard to hold on in their shifting ground that they never reach a climax state. They are pioneers. Such a place is open, as all earth's shores must be, to drifters, like the black stork.

The driftwood that lands on the beach and sometimes piles up in great numbers and bulk on the upper tide level after a storm, could come in from almost anywhere: Africa, Brazil, Massachusetts, Maine, or Nova Scotia, depending on how it was transported, by ships or by the sea itself. Years ago, sailing ships traveling along the Outer Cape with cargoes of lumber chained to their decks might encounter heavy seas and be in serious danger of grounding on the shoals, in which case they would occasionally jettison the cargo, which would land up and down the beaches, to be picked up by those famous human scavengers, the "mooncussers." Since such lumber was often in the form of planks or studding, it supplied many a family with material for their houses. I can think of at least one house which is largely constructed of it.

Or as it happened not so many years ago, a log jam in a Maine river broke the boom and the logs went careening and dipping down to the sea, a great many landing after a while on the Outer

Beach. Huge trunks of trees sometimes appear, carried in by the sea. I have found cherry, red and white pine, cedar, spruce, beech, and even some canoe birch with the bark still on it, a tree not indigenous to the Cape. Mahogany and walnut have been found at times, and a few years ago the cross section of a tree was discovered near Eastham that turned out to be a very hard and heavy wood from Brazil, probably fallen off a ship. Parts of dories, or larger vessels, broken oars; buoys of all colors and shapes, glass floats from lobster pots, branches, logs; boards of many different sizes and lengths, wharf pilings and planks, and dunnage, timbers used in stowing ship's cargoes, cases of scotch, always, in my sad experience, without the scotch; crates from vessels of all the world, South American, Russian, Japanese, French, and most of the nations you can name; all these and more have been carried by the sea, sometimes for twenty or thirty years, until they were finally landed on the beach. It is wood for the fire, a house, a shack, or a table, and material for any curious scavenger, on behalf of aesthetics, science, or history.

The driftwood is a migrant, to move again soon, unless it is taken off the beach, burned in a fire, or lodged and buried deep above the high-tide line. It may serve temporarily as a place where seaweed and other litter gathers, or where crustaceans might congregate. The birds, if it is an accessible clump of branches fingering over the sands, rather than a log or heavy timber, may peck through it after such tiny animals, their tracks making a delicate tracery running under it and arrowing away. Driftwood migrates like the sand and the birds. It is another aspect of the surf's swing and draw, its dragging out, its removal and its deposition, part of the constant remolding of this shore.

On the cliff tops too, over the beach and the round horizon, everything goes out and round and returns. A curve is the only rule. As it does everywhere on the Cape, the wind goes across from one direction of the compass or another, streaming with light and moisture, lifts up, lifts you to it, and with long low swoops, sudden breaths and seething, it whisks the waters of the marshes and inlets, rounds their brown shoulders, races through trees and over cliffs clean through across the sea. The land under it, held down more definitely than beach or dunes, also waves as they do.

The heights above the beach, the low dipping slopes and hills, though vulnerable over long periods of time, foot by foot and yard by yard, look unrelievedly intense and bold. They glisten under the open light, the open draws of the sky. There are miles of scrub oak, bayberry, and beach-plum thickets shining as if they were wet with light, or, in the winter months, purple, maroon, and diffused with blue like a mist. This is where the fox and song sparrows gather, and the myrtle warblers. There is a sound of leaf ticking and branches tapping together above the pouring of the surf.

Sandy tracks made by beach buggies claw through wide patches of huckleberry, which have red or bronze leaves and conspicuous red tips to their branches in the fall, and in other areas the ground is held by beach grass and sometimes wide mats of shining bearberry, or hog cranberry, green and purple with bright-red berries under their leaves. Wide patches and hollows of blown sand are growing with Hudsonia, "beach heath" or "beach heather," which is a soft gray green, and has golden yellow flowers, changing to darkening gold before they die, flowers, incidentally, which have a faint but sweet scent to them. Sometimes they are accompanied by "reindeer moss," that seems to hold on tenuously, since its gray-green fronds crumble up and blow away, though in point of fact each of these fragments can lodge again in some other area. In the grayest of weather this lichen seems almost luminous, having a sea shine in the rain.

Piny hollows circle behind this spare vegetation, the trees with burnt-orange leaders killed by salt spray, and oaks, often dead at the top, along with a great range of scrub; and until recently when building was curbed by the National Park, new clumps of cottages and half-finished roads appearing all the time in new areas.

The cliff-top landscape is irregular, tilting up and down, dipping back as a rule toward the west but in varied planes. Just above the beach its hollows are scoured out by the wind, almost denuded of vegetation, deep cups with drops below them sheer down to the beach. I have seen the remnants of house foundations in such hollows, or a creosoted pole or two sticking up above the surface of the sand, not too old by the look of them, proving what an ephemeral habitation such a place can be. Where the low growth holds on, sometimes in masses, like bearberry, or in patches like the Hudsonia, it too lacks a certain finality, giving a free, waving look to the surface of things. On the other hand this vegetation is definite enough. There is no fragility to it. It is scraggy and tough. The strong shrubby growth may be held down but it also gives the landscape a symmetry and economy; it does not give the impression of being hit or miss at all but very definite and sure of its place, as sure as wind-struck, salt-sprayed plants can be. Each plant stays rooted from place to place through this sandy earth, being adapted to intense light, drought, and constant winds, holding on hard against being scoured out and displaced, and ready also, to move into new areas. Beach grass, especially, has this ability to move in on newly deposited sand, or where "blow outs" have occurred, areas in which the wind has finally blown the sand out from under the plants formerly rooted there.

So this patchy, heathlike region is held down in substance, temporarily, if not in form, adapted to the constant changes made by the wind. Closer to the cliff's edge there are likely to be hummocks or mounds, like those of the dunes. A high hummock may be held down by beach grass and have a core of bayberry bushes with only an inch or two of leaves and branches sticking out at the tops. Beach grass, bayberry, seaside goldenrod live in close if embattled communities, at least with respect to the wind. These plants and others may all join in holding such hummocks or mounds together, while the Hudsonia in rounded clumps holds and extends its grounds across the level sand around them.

There are two principal species of Hudsonia by the way, ericoides and tomentosa. Both have been called "poverty grass," but the name is usually applied to tomentosa, which is the more common of the two. They are not always easy to tell apart. The ericoides, sometimes called golden heather, has tiny spinelike leaves that stand out fairly distinctly from the stem and each other and it is a plant that stays green for a much longer time during fall and winter. The tomentosa is densely tufted, downy, softer in appearance, and it turns gray, or bluish green, being subject to winter kill more readily than the other species. On Nantucket at least this plant used to be gathered, dried, and used for fuel.

The Hudsonia are "xerophytes," plants that are adapted to extremely dry conditions. Their tiny leaves offer a reduced surface in the face of intense sunlight and therefore do not lose water so readily. A "succulent" like the seaside goldenrod, on the other hand, has large fleshy leaves for storing moisture, another adaptation to drought conditions. This region is no desert. Even the term semidesert has to be used with caution. Its annual rainfall is the same as the rest of the Cape, but it is relatively unprotected and lacks the topsoil needed for the plants and trees not adapted to it to send down roots fast and deep enough to get moisture. The beach heather, stem-rooted like the beach grass, probably evolved in an alpine environment, where conditions were considerably worse than they are on Cape Cod at present, and moved in to the Cape during the postglacial period, remaining ever since.

Still, the unprotected, dry ground is eloquent enough of the assault made upon it, and the eroding cliffs with the plants that hold down the ground above them become part of the fierce sweep of time and oceanic weather. Here is a lesson in exaction. Perhaps those omnipresent Cape trees, the pitch pines, show the hard effects of a sea-edge environment more obviously than most. They cannot survive too close to salt water, but a little farther back the results of wind and salt spray is to kill their leaders on the windward side, dwarf them so that they grow flat on the ground like the Hudsonia, or to tie them in knots.

Everything has its method of survival. Each gradation of the ground, each hollow, slope, or level area, has a life to fit it or to visit it. The plants move forward seeking water. The birds fly through the thickets hunting seeds or insects. The exaction lies in a frame of reference. There is a quality of trial by the seashore, of odds, which taken care of by a mere plant, seem no less formidable. Their success in coping with the situation within its limits and precise needs is allied to all life's insistence on success.

We put great emphasis on the flowering parts of a plant, and certainly the golden, summer-yellow of the Hudsonia, growing in bunches like bouquets, is rare and beautiful over the bare ground with the blue sea stretching beyond; but this plant is also rare in its restraint. Its tuftlike branches, its leaves, spiny scalelike or coarse textured as they may be, have a beauty, a resourcefulness which is the end result of ages past human knowledge of them. They are a successful experiment in creation, artfully finished and well related to the world.

XII

The Depths of Sight

Where is that eye to the sea beach and the sea that I might enter, to follow further than I know? There are so many unfinished depths suggested by the surface of things. A wet, white and gray pebble of quartz has the kind of grain that leads off to snow and rain and all the watery and windy associations of earth history. A feather, fitted, barbuled, light and strong, holding the air, refracting the rays of the sun and using them for its colors, has the horizon's curve and the graces of the sky. The bryozoans on the seaweed tell a deep and primitive tale about the salt water and its animation. We should not be so impressed by our powers of assessment as to take things merely at their face value. To see more than the outside shell of the landscape we should be ready to admit its depths and whatever takes part in them, admitting also, that we are limited in our own capacity.

It is not necessarily what I see as I walk the beach that might make sense to the world but what sees me, even though it can't write a book or drive a car. In the eyes of birds for example is a special kind of closeness to truths of nature which we might only see through a glass. Their very distance from us seems to prove it. Look at a herring gull and you see an animal with less intelligence than a goat, but with the same ungiving topaz eyes. I kept a female brown thrasher once for a week or two and there was nothing her sharp yellow gaze had for me but a constant glare, perhaps nervous or agitated but not to be deciphered otherwise. Consider the eyes of an alligator. They are not even revealing enough to be called "enigmatic," which might be a misleading word in any case, implying some half-human wisdom like a sphinx. Its eyes are mere sunlight openings, cracks, and crevices. Its lids are turrets, drawn down on a bit of nameless colored water. Other animals, other societies, receive natural messages in ways that may have no more excitement in them than the reflection of a cloud passing across the surface of a pond, and still they may know what we do not, and the place they live in and respond to is our envy to discover. The strict, close relationships in the world of life, the life of earth, result in sensitivities which are no less rare for being divorced from self-knowledge.

That scavenger the herring gull may be just as lazy as it looks. Human civilization has done nicely by it. It can live off the "produce" of our dumps during the wintertime, when it would otherwise have to work for a living. When a gull is standing around on the beach looking as if it were doing nothing, and we ask why, imagining the same specific purposes we think we ourselves pursue, we might be disappointed. As likely as not, the gull is doing just that, nothing, and will fly off at some stimulus —hunger, another gull, a plane, a man, or a shadow. And yet it is the bird's association with the seashore, its response to the currents of the air, to changes in tides and weather, to the sun's appearance at dawn and the departing light of evening, that lies in its own sight. It is just possible that you cannot exaggerate the effect of light on the physiology and actions of a bird. At least it seems to be of primary importance in the cycles of migration. So in a herring gull's cold eye is a receptiveness not so much qualified by intelligence or the lack of it, but inextricably, directly connected with the world of light. When birds and animals react to me, and why leave out any man or child, even if it is only in answer to an "escape mechanism," I see a vision unexplored.

One morning several hundred gulls, herring and blackback, were congregated far below and ahead of me as I walked along the cliff. As soon as I appeared on the edge, casting a shadow over the beach, they took wing, even though I was at least a quarter of a mile away, and they rose in one heavy flock and beat slowly away down the sands and the surf line.

Not long afterwards I saw an Atlantic, or red-throated, loon swimming just offshore, tall necked, its head looking off and alert as though the bird, like a pilot in his house or a watcher at the masthead, was on a constant lookout. When it saw me it glanced wildly and ducked head first, over and down, slipping under the water.

On the same day, a few miles further along, I saw two harbor seals of good size, swimming twenty or thirty feet outside the beach. First one dark head appeared above the water and then as I watched through field glasses from the cliff top two big dark eyes suddenly looked up at me, and the seal dove, followed by another one a few yards behind. The two swam through green rolling waters parallel to the beach, coming up every half minute or so, their swimming forms like shadows slipping through the sea. The harbor seals, though intelligent and appealing animals, have suffered great persecution by man and are much less numerous in Cape waters than they used to be, so that the sight of these two large specimens at home and roaming along the shore was a great pleasure to me, and above all I enjoyed having made some contact with them, as I did with the birds—the mutual life touched on, an electric communication made between one far pole and another.

Sight in our sense of the term involves symbols in a very special way, but it is part of a universal trial of knowing and reception, and in animals without consciousness and means of assessment this may mean more than automatic reaction to light. I think of a crowd of newly hatched minnows like tiny slivers of glass, running up and quivering through the water. The most definite thing about them is their large black eyes, contrasted with a bodily transparency so fragile as to seem past fragility, an artifice of growth on balance, in a chain of universal actions that might have their matrix in a dream. Those large eyes are the eyes of first attainment. Sight is the expression of an alliance with the world in lives twitching and quivering toward mutual attachment and effect. It may be the gift of misery or adoration in a man. It is the opening of gates in a child or an animal new to life.

Perhaps when you look at, if not in to, a fish's eyes you are looking at depths of water, an animate fluidity. In its senses there is a watery knowledge with a supremacy of its own. What a lightning and at the same time a listlessness there is in them, in their hurrying ways through currents of fluid light, and their suspension in its stillness! Many of them only last for a day or a few minutes before disappearing as a food for other animals, in the mercurial depths of water allied with life, this intoxicant, this terror.

My sight meeting that of a gull or seal crosses and contains this landscape, environment, or place of existence with its own eye and its own depth to find. The expression of water, sand, and sky leads vision beyond itself.

One quiet, moderately cold night when the mist hung so low over the water and beach that they were closed in, but at the same time illuminated by the moon, I saw the port and starboard light of a fishing boat that looked to be only a hundred yards or so away down the shore. I kept walking toward them with the illusion that the boat was moored close to the beach, but after a couple of miles the lights were still receding and I turned back.

The tide was close in and sheets of foam pitched in and dragged back with a sound of rattling stones but in gentle rhythm. It was a quiet sea, and beyond the surf I could detect little strikes of light, the curling over and stirring of white and silver. Up through mists and wisps of cloud the moon appeared intermittently, riding above the water. The beach was covered with soft airs, its distances diffused in gray and pink and pearl, a mood of ambiguity. I felt that whatever I might hear or meet up with was out of my control, at the dispensation of the world in and beyond the atmosphere, having unknown connections light years away and joined with fish and moon and speeding globe. In this isolation, a familiar place turned inexact and mysterious, I felt I might sense all sorts of far nerve ends tingling out of the night behind the mist. We receive very little of what reaches us out of this tribal universe, whose messages light through us unseen and unheard until we, as individuals, are turned to the dust of the sky.

Night or day, the sea and sea beach offer their changing spaces of light. One afternoon in January, halfway between hours of warmth and hours of cold, rain and snow, morning and evening, the sea off Nauset was racing green, spray tossing off the tops of the waves that simultaneously paused, curved up, and broke down in thunder. The whole sky was full of cloud featherings borne over before the wind and along the horizon out to sea were colors of lavender and gray, and pale-green openings like caves. The wide, steep beach was full of gloss, with a roll and fire of its own, and above it fringing the edge of the sandbanks the beach grass curved out and waved. I felt a resonance in the beach, a tremendously heavy and vibrant tone, the tonnage of sand and surf in harmony along with a low moan from the sea's lungs.

Small flocks of black ducks quivered over the water and then flew in to Nauset marsh. Then the heavier Canada geese beat in with stalwart wings, to thin out from their V formation to a long line as they wheeled in low against the wind and then regathered as they settled down on the marsh.

Blackback gulls glided low across the outer line of the surf and sometimes their shadows appeared on the curving wall of a wave. Herring gulls soared in the heights and then beat forward on sinewy wings like flounders pulsing and beating through the water.

One gull flew down the beach with a ribbed mussel which it had found on the marsh and dropped it from high in the air. Then the bird retrieved its food and tried again, taking a chance on whether or not it would strike some boulders and break, since this is a haphazard and not a very knowledgeable game with the gulls. They pick up the habit from each other, by example rather than inheritance. Sometimes it works and just as often it does not.

The seaways of soft feldspar green foamed and flew, and the clouds ran. Thin black strings of seaweed lodged in the sand were waggling back and forth in the wind. There was a swish of milky surf up the beach. Over the uncountable numbers of sand grains, each with its own size, shape, and color was a clean radiance, even a magic. Because in this realm of wide, majestic use, of continual advent, each offering was still of a proportion perfect for its moment in time. Each single action, the silhouette of the straw-colored grasses curving before the wind, or a gull shadow on a wave, a crystal grain sparkling in the light, was of such an excellence as to defy category or name. And they were magic and miracle in their shape and ways of use because they had life's inveterate sanction, and that above all else is not subject to lessening or degradation in this world of nature.

Like the lights that appear under the mist, or over the open barrens of the sea at night, like St. Elmo's fire on the *Pequod's* mast, there are electric tricks playing on the horizon, perhaps at all times, since there seems to be no end to light's action over the waters with the sky's depth behind it. As I walked up the beach there was a radiant white patch hanging up in the soft, scudding overcast, not in the sun's direction—reflected off the water perhaps—but having a wild aura of its own. It gave me a feeling of communication with something which had a right to awe. We may have passed the primitive stage, but the primitive respect for what was beyond human control and the magic used to propitiate it or bring it to play may still have their sources. The light and its manifestations is still too quick for the eye, or for the facts.

Science itself goes on proving that there is no infinite exactitude and that many things can only be explained in terms of probability. The fact that nothing is stopped by our constant search for a simple solution to life is what keeps science in business. The search into the nature of cells finds them full of inner whirlings, the motion of countless component parts, of a universal restlessness. They are structurally fantastic and each kind is manifestly different. Our voyage toward the invisible is unending. The molecule or the jellyfish, seen through one human lens or another, retain their share of the marvelous. And if we marvel, we are still capable of learning.

A radiance above me, a changing freshness in the air, between warm and cold, a shudder of wings over the beach, another language of unexplored dimensions, life expressions understood in terms of sight and spirit, and still to be learned—the nonhuman advents that pass the limits of a man. There is a common realm of action and perception, whose boundaries we may never reach, where men can be more grateful for their belonging than their isolation. It is part of the changing state of inanimate things, the response of lesser forms of life to the construction and motion of the world that invades them and which they invade, and it is acted out by the mind. The tidal waves run through us all.

To see as men see and merely to react like a moon snail or a horseshoe crab to the difference between light and dark are two representative actions in the same vast realm of response.

Do men belabor the special nature of consciousness too much, as if it were some kind of A-1 badge that separated mankind from the rest of animate creation? Consciousness must be infinitely more mysterious, more connective, than any attributes we may assign it of personal distinction.

XIII

The Flight of Birds

The appearance of migratory birds in fall and spring, or simply their constant activity, suggests their range. The ability that a gull displays in the turmoil of the air is enough to bring other winds to these shores, to make you realize that the beach joins the long shore line between Cape Cod and Florida, that the waters to the north of us move on toward Labrador and Baffin Bay. Their wings are allied to the circulation of the North Atlantic. New England is not so far from the Arctic Circle, and when the auks, the old squaws, or the buntings come down to Cape Cod in the autumn they bring the proof with them.

We have had an appalling record this side of the Atlantic, of decimating the population of sea birds, which are more vulnerable than other species because of their nesting habits, on islands or rocky foreshores. The great auk has gone, and the puffins reduced to small numbers. If we were able to kill them all off, either on purpose or through lack of responsibility, what little island people it would make us!

The very colors of a murre, or a razor-billed auk, a contrasting black and white like penguins, suggest the black cliffs and rocky headlands where they evolved, the white snow and ice, the cast of deep and icy waters. One June day, when the beach at Race Point was glaring with light, and all the winter leavings, like the twisted dead stalks of dusty miller, were being replaced by a freshness in the shine and scent of things, I saw a dovekie, or little auk, on the beach a few yards away from the water. It is a very small bird, though conspicuous enough with its penguinlike stance, its black and white plumage, and though it was in full view of a number of bathers no one saw it. When I approached, this seasonal anachronism ran rather than flew away from me down the sands into the water where it promptly dove out of sight to bob up out of harm's way many yards offshore. Since most dovekies return north in late winter, I supposed it was a "nonbreeding straggler." They migrate south in the fall to more temperate waters not locked in ice like their home feeding grounds. Over a period of years and at unpredictable times, there are "Dovekie wrecks" when these birds are blown inland by gale winds and show up in the most unlikely places: ponds, back yards, side roads, gardens, filling stations or shopping centers. Since they are not able to take off from land with any ease, if at all, they are vulnerable to predators of all kinds, provided they survive exhaustion and starvation. Some years ago I saw a number of them lying dead for several miles along the Cape Cod highway.

The dovekies are messengers from the north. The way the gulls use the wind as it is deflected from the waves, or ride into it, hovering, then gliding down, is symptomatic of the sailing skill of other birds that travel far beyond the shore, the aerodynamics of the open sea. They are masters of the art of air as no plane can ever be. I remember watching some fulmars in the wake of a ship one wind-tossed day, the great blue-green waves in rocking fullness shouldered with foam. They glided between the crests and troughs of the waves with effortless deliberation, and then lifted, curved away in a wide arc, and returned. Back and forth, they seemed to tip the water's surface with their wings and clip the waves, gliding and curving with them, expending no

excess energy at all. I felt them rise on the upward air in my lungs, my admiration.

In birds you see pure action personified, an endless spontaneity reacting to the air, the season, the light, and on clear nights the constellations that may help them find their way. A flock of red-backed sandpipers or sanderlings, all spinning, wheeling, and sun-reflecting at once, have an ecstatic dash, a common brightness set going in them which must carry them a long way. They are long-distance migrants flitting from one end of the earth, one shore line to the next, and judging by their actions it is hard to believe that they could ever rest. Searching for crustaceans or sand worms along the beach, they run on flickering black legs, bodies tilted forward, flitting, bobbing in syncopation. When close to the surf they may fly up briefly when it piles in and then drop down again when it retreats. With their quick, automatic run, and heads constantly jerking forward and back they seem to be endowed with an almost comic gift of hurrying.

Suddenly, with a sharp piping cry a sanderling flies off the beach and then disappears like a gray chip over the water, a tide bird faster than the tides, where there is no following it. This bird is quick and sweet, and cleans the earth of too much hesitation.

Of all the birds that visit the beach during fall and winter I take most delight in the snow buntings. They have such freshness in them, skimming the cliffs, rushing by like bits of foam. The white in their plumage is so pure, snow paths between markings of black and cinnamon, like briers and weed stalks, with suggestions of greenish gray when the sun shines on them. They are birds of the Arctic tundra, companions of the musk ox. They fly up suddenly, as they are constantly doing at the least disturbance, their whiteness dancing up above the beach or along the faces of the cliffs, and then settle down again, pecking away, at home in wastes and barren land, the lonely stretches of the world, these are flowers, snowflakes, foam, fitted to a poverty and its freedom.

They are seed eaters like sparrows, and may also eat such tiny creatures as they find along the beach, and they are always flocking and scattering out from one rise and level to the next. To me, the fanciful difference between buntings and sparrows, sanderlings, gulls, horned larks, and many other visitors to seaside lands is their trait of invisibility. It is not only their whiteness—they look almost entirely white seen from underneath, appearing and disappearing like clouds—and a plumage which belongs to the accents of sunlight, grass stalks, dune shadows, on the bare ground—but their actions. With a motion reminiscent of the roller-coaster type of flight which the goldfinches have, flocks of buntings will pour down onto the cliff top or beach, spread out and then fly up again, with an inner billowing, a dipping, and rising as they go. Twittering with a note of tinkling bells in the high air beside the bowling sea, they swing and then burst in gentle snow flights across the ground, through one opening, one neat run, one clean escape to another. They turn the invisible into reality. They have a continual lift, the agitation inherent in all life. They fly up ahead of me as sparks out of the unseen rest and center of things.

Another bird of the tundra, a specter from the far north which appears irregularly over the years during wintertime to hunt for rodents and occasional birds along the coast is the snowy owl. I remember seeing a mounted specimen when I was a boy and thinking it was the most desirable thing on earth to own, and since I never did own one, the snowy owl stayed intangible and magnificent in my mind; and the first live one I ever saw did nothing to disabuse me of my impression. They migrate to beaches, salt marshes, and islands along the coast, choosing elevations as a rule, hummocks, knolls, or dunes from which they can survey the surrounding countryside during their hunting season, watching the man or beach buggy arrive as well as evidence of prey. The one I saw was way down the south end of North Beach, that stretch of Nauset beach which ends at the straits separating it from Monomoy. It was perched on a hummock, and at first was nearly indistinguishable from the top of a white picket fence buried in sand, or the kind of white marble marker, rounded at the top, which you might see on a roadside in Vermont. We were driving toward it in a beach buggy and when it flew off low with big, soft, bowed wings, its feathers, white and flecked with gray, took on a blue-ash hue from the winter light and the uneven shadowy land around it. The great owl lighted calmly on another hummock further on. It stared straight at us out of fierce yellow eyes, with inscrutable dignity, and when we turned and came at it from another direction its head almost swiveled all the way around, looking at us from over its back. It kept its place in center stage.

Many thousands of eider ducks winter in Cape Cod waters. During October and November especially they can be seen shuttling back and forth across the sea beyond the Outer Beach. Some feed, principally on mussels, in the bay region or off Chatham and along other shallow shores and inlets, but the majority—an estimated 500,000—spend the winter over the shoals between Monomoy and Nantucket. Seen close to, as they fly low over the water, they are as sturdy, clean shaped, and of good design—the red-brown females, and males patterned in black and white—as a coastal vessel, a dory, or a skiff. From the beach you can see them fly over water in single lines, sometimes as much as a half a mile or more in length, with a steady, throbbing flight, like a suspended string of beads, alternately white and brown.

By contrast brant fly in longer, thicker lines, and sometimes show up like shivering black specks high over the sea. Well into December the gannets pass by over the sea surfaces too, flying singly for the most part, their broad white backs and long black-tipped wings reflecting the sunlight as they turn, to dive in their grand manner down, from fifty feet or more in the air, hard and bold into the water, sending up jets of spray.

Clutching at any aspect of nature is to seize a drop of water in your hand. Ebb and flow passes the great beach, the eternally wide ebb and flow of day and night passes the cliff tops, all earth's shadows wave across its seas, and yet this is the precise route of the birds, their direction and their home. They know its guidelines inwardly. For us, who put so much emphasis on outward instruments, this can be almost impossible to understand.

Still, we can exaggerate the division between us. We are all at home together, however we use the stars and seasons in our separate ways. Men are as subject to mortality as birds, even though the latter can't dwell upon it. They in turn are vulnerable to chance, to disease, to going astray and meeting with mishaps when confronted by the freakishness and violence of the weather. Many a duck or sea bird, caught on a lee shore or in a marshy inlet during a great storm may be unable to rise into the wind and is exhausted or swept away and seriously injured while trying. Life and death, joy and disaster, go wing to wing. Birds have less capacity to deceive themselves than we, being unable to avoid the perils of nature and at the same time its protective power.

I had similar thoughts in mind one day in November during a violent coastal storm while watching some gulls, ringed-bill and herring, together with a few shore birds, that were gathered at the head of an inlet along a relatively sheltered part of the Bay. The Outer Beach was of a violence that day which could hardly be approached, either on foot or in contemplation. Even here the storm winds were relentless, hard and cold, flicking and driving the sands along the shore, whipping the marsh waters behind it into a froth. Sanderlings made short, low, flying hops back and forth, but were unable to do their usual free hurrying and basket-swinging flights along the shore. The gulls stood in shallow water facing the wind, water that was being whipped and lashed, and sometimes they would drop down sideways a little before the wind's force, thrown slightly off balance, acting like a man who has been cut across the face. Taking to the air just above the ground they would find difficulty in maneuvering and were forced back, sometimes fifty feet or more, to continue standing where they dropped back to the ground; but even in this they showed a certain supple power, a control aware of its limits, the sinewy economy of wings lifted in the wind, the plain sky beauty of feathers gray and white. The storm was ending, although the water was still being whiplashed into foam. The light was very cold and the sky line was heaped with sunset fires.

Surely everything, everywhere, was vulnerable, and yet it was that bird closeness to such primal powers as might seem to us bitter, alien, and cruel—the gods of the north, of the waters and the winds—that gave them an essential balance, a rightful place. That great sky of theirs was unexplored. It came down to me that regardless of what he learns, there is so much for a man to go on asking.

What can birds tell you, other than displaying those traits of aggression, or fear, or mutual attraction, which we may recognize when observing their behavior? We have a little fear in ourselves, when looking on, that we may go too far in mixing up our own traits and terms with theirs; but each will manage to keep his territory, untransgressed by the other, and each takes part in the high order of nature. Watching the birds, I have seen ceremony, ritual, love-making, display, all worthy of admiration by the most glittering of human cultures. The speech of men and the speech of birds do not divide us altogether. In silence is unity.

Perhaps the most eloquent thing about birds is that which we will probably never learn to decipher. In his study of puffins, R. M. Lockeley refers to their "subtle, silent-gesture language."

That language is part of a still more silent order, the dark realm of existence where all their actions and necessities have their play. Approach with patience and with care.

One day I had walked for several miles along the cliffs toward Eastham, through thickets of scrub oak, and bayberry that smelled very pungently in the fall of the year. The sky was full of shifting winds and the day as I walked full of weather changes, from an edge of cold to warmth and back again. An early sun began to be covered by pale-gray clouds and there was a mauve light over the sea. I caught sight of a little wren along the way, and there was a number of sparrows, both seen and heard—song, chipping, seaside, and probably others. It was a low, shifting thicket world full of potential surprise, bordered by oceanic sound, rocking with light and air.

I retraced my steps a few hours later over a narrow sandy road, at times no more than a track, and I saw a pigeon hawk flying off ahead of me, stroking deliberately and quickly with its long wings. Then I noticed another one roosting on a broken-off tree several hundred feet back of the cliff just outside a wood of pitch pines. The first one made off in that direction too, roosting not far from its companion on a dead stump, and they both stayed absolutely still, like falcons on an Egyptian frieze. I could hear a blue jay screaming somewhere in the background.

I noticed feathers scattered on the path, gray and blue, blowing ahead of me; and then, there it was, a blue jay freshly killed, its breast bare of feathers and shining red like some rock wet with sea splash in the crimson path of the setting sun. What kind of a game led up to this? Could the two hawks, one tempting the jay by its distance, the other scaring it by its proximity, have managed to send it out into the open where it had no chance against their swift and effortless pursuit? I walked ahead for a short distance and then waited, watching through field glasses for the hawks to come back. The nearest one did, after a few minutes, beating down tentatively over the kill, then rising again and leaving with its supple flight. The other had moved a little closer and roosted on an abandoned telephone pole, full of an ancient poise, wonderfully still. After that, I am sure, they never went back to the road until I had gone for good. The grace and tension, the space in that formal scene stayed with me for a long time.

XIV

The Marsh

The Outer Beach is broken only at Nauset Inlet, where the tidal waters pour through an opening that has frequently changed its width and position, and at Chatham. The Chatham break leads in to the wide area of Chatham Harbor and Pleasant Bay. In both places, but more especially at Nauset, where the marshes and the inland shore behind them are protected by the beach and a sandspit some two and three-quarter miles in length, an unstable, but at the same time fairly constant equilibrium is attained between sea and land. It does not seem obvious that this should be so at all. The sandspit looks only too narrow and fragile, and at intervals it does show evidence that the sea has broken through. Driftwood logs lie on the cuts made between its hummocks, headed as they were when the sea subsided, after it had lifted them in toward the marsh.

Except for the great volume of the beach itself, which is maintained in collaboration with the forces of the sea, it is hard at first to understand why the marsh should not be inundated. Why does that lord the sea not heave in and overwhelm this sandy barrier, flooding over the marshy flats and islands, and wash up permanently against the inland shore?

The shoulders of the low cedar-studded land slope down to the edge of the marsh with a neat, trimmed look and neat houses, seemingly confident of being in residence indefinitely, although I have heard people who live there talking in ways that suggested they were not sure of it. Once see those stormy waters heaving and rushing over the sandspit and you cannot be sure of anything. Looking out at the sea, even from a fairly safe distance, you can find eternal balance and at the same time inundation and disaster. Now that the Outer Beach stretches past the miles of cliffs and is no longer backed up by them, becoming an outlying stretch of sand, its own "protective" power might seem much less clear. On the other hand, when was this beach in anything but a state of flux and change? There is protection in that, even if it is hard to define. The fact is that the relationship between the sea, the beach, and the sandspit, the marsh and inland shore, has been maintained for ages in the past and probably ages to come. In general the volume of sand that is packed along the shore balances what is removed from it, but only in general, for the time being, because erosion takes place consistently over the years and during its course more sand is removed than delivered. Also a standing equilibrium is kept between this deposition and taking away of sand and the conditions offshore: the currents, drift, wave height and direction, the changing shoals and bars. All these states and forces are involved in an extremely complex kind of order, and it is certainly broken and rearranged all the time. A season may show it, or the records of history. In fact, changes occur from day to day.

When the young explorer Champlain visited the Cape in 1605 he sailed into Nauset Harbor, and at that time, judging by old records, the inlet was about halfway down the sandspit behind the beach. Since then it apparently has moved about a mile south, but its entrances have changed now and then, with long periods of relative stability in between, which might be broken at any time and then followed by some new arrangement of forces.

In his *Birds of Massachusetts and Other New England States,*

Edward Howe Forbush pointed out that this long protective spit, or "beach ridge" extending from Nauset to Monomoy had been pushed back a considerable distance, perhaps a mile, since the early seventeenth century. It used to lie far to the eastward, judging by early charts, of where it is now, and took the form of a long narrow island some twelve miles in length "with several small islands north of it and outlets to the ocean at either end— the northern one at Eastham and the southern lying between the end of this beach ridge and the Chatham shore.

"In 1854 during the great storm that wrecked the lighthouse on Minot's Ledge, the sea broke through the barrier into Orleans water at Nauset, and afterwards much of Nauset Harbor near the entrance filled partially with shifting sands."

The recent Woods Hole beach studies report that: "The spits literally broke into pieces and the inlet itself became quite complex in 1957. Nauset Inlet has done this before. A study of coastal charts shows that Nauset Inlet opened hard against the cliffs on the south side from 1856 (the first good chart available to us) until 1940. Charts of 1941 show that in a single year a spit grew from south to north against the littoral drift and shifted the inlet a mile to the north."

For some length of time, the storms of 1956 and 1957 resulted in two entrances along the spit, one of which closed up subsequently. Other temporary break-throughs can be seen along the spit, varying from 150 feet to a few yards across, extending down its length until it joins a broad, high stretch—almost a long mount —of sand which ends at the present inlet, with North Beach on the other side. This sand is subject to storm flooding and to winds, to being removed and added to, recut and carved by the waves, and except on the marsh edge of it, beach grass is not able to gain a foothold. In recent years four or five hundred pairs of terns have nested there, and are protected.

The volume of this sand is immense. It shelves down steeply

toward the water where it becomes part of the beach; and where the channel of the inlet curves in, the ends of the beach on both sides keep changing their lengths and relative position. The sea builds high shoals off and around the incoming tidal channel during one season and it may level at least parts of them off in another. During the summer of 1962 the ribs and bottom of a boat at least thirty feet long was revealed on one bank of the inlet at its mouth, and could be seen for months; but by the winter of 1962–63 it had completely disappeared. A sandbank lay over it which was at least five or six feet higher than sea level.

Aerial photographs taken when the spit broke up in 1957, and afterwards in 1958, show a very elaborate and confusing pattern. Shoals and separate spits began to drift, to join and separate, shift and intermingle in curling, curving folds, an interwaving and repositioning of sand materials that would seem to have no parallel in nature.

The Nauset Inlet is being driven into the marsh behind it at an average rate of about 2.8 feet a year, except in years of extreme erosion. This figure is about the same as that of the cliffs, and on the whole it is probably somewhat less here than there, although the marsh area is being very gradually diminished in extent. Its wide channels and bays, its marshy edges, islands, and flats, are held in the balance of great forces sweeping along the shore, or occasionally breaking through in violence. Although it absorbs and releases the tidal waters with ancient calm, it seems wide out, subject to the sea and a part of the complex, barely understood forces that build and break along the shore.

The marsh is a refuge for ducks and geese, and gunners for centuries have waited there for the "whistlers," or goldeneyes, and the black ducks to whir, swing in, and career overhead under the wide light of dawn while the cold wind ruffled the open water and stirred the matted grass. Like the tides that flood in and fall, like the marsh grasses that grow and wave, then die down and take on their matted winter look, or the marine animals that swim in through the tidal channel and go out again to sea, it is a place of flight and motion. The local animals, crabs, clams, mussels, snails, the salt-water minnows in the ditches, the marsh snails, and numerous others, must go through their cycles of growth and death and decay here, the building of interlife relationships, but the over-all feeling that I have had about the marsh is a certain bare economy, as though it was more obligated to migrant forces, to flooding in and flooding out, then to any enclosed stability of its own. In a way it has the wide, flat isolated look of the more sheltered and extensive marshes on the Bay shore, but it is an isolation bound to the open waters of the sea which run through it and sometimes threaten its borders.

After their green summer and early golden fall, the marsh plants and grasses darken. In November the marshes are still russet, umber, and yellow green, but by January they are dark brown with reddish tawny tones in matted grasses having the coarse texture of a deer's coat. The saltwort plants, so fresh and green and full of salt juices in the summer, have turned dry and white, curled over at their tips so that they have the look of singed wool.

When you walk behind the sandspit the marsh flats seem to stretch far off toward the shore and the channels between them are partly hidden. Nauset from the landward side, on the other hand, looks as if it were mostly composed of water, especially at high tide. It is both a good country for low-grass lovers like sparrows and those that ride its watery lanes and lakes like ducks and geese.

Low-flying, drab little seaside sparrows fly up off the grass for short distances and then disappear again. Occasionally I have flushed a meadowlark that planed up over the marsh. Horned larks peck in the dunes, tripping forward with a stamping motion of their legs, and then stop, to stand with a backward slant to their bodies. They fly up suddenly with shrill lisping cries; and all the while the deep quacking of black ducks sounds from far out in the middle of the marsh. There are always gulls, far or near, with their slow gliders' fall and rise on the wind. The great blackbacks fly heavily overhead, sometimes wheeling in circles over the inlet with a muted baying, or hoarse, guttural calls; and with their necks and heads stretched way out and their wide-spread wings they might be mistaken for gannets.

Red-breasted mergansers come in from the sea with their thin heads and bills straight forward so that in flight they become throbbing arrows sent from a bow. One evening I stood in the hummocks of the spit facing the marsh while flock after flock of Canada geese flew in overhead, bugling as they came, close enough so that I could hear the fine high whistling of their wings, and even a rattle and rasp of air through their feathers. Low-flying planes often start them up as they feed in the marsh, along with the wary black ducks, whose cloudlike flocks stray back and forth for a while before they settle down again. A black duck's wings show white underneath and they seem to spin as it flies up high and fast and changes direction, like a weathervane.

Quivering, soaring, swinging flights set out over the wide marsh, and the bird fleets ride the waters. The goldeneyes follow one another bobbing along in a channel, along with mergansers and occasional buffleheads, whose white heads or sides suddenly shine out as they round a corner. A rush and glide of water shows brightly in the distance when an eider plows quickly forward. The Canada geese feed over the marsh or on the borders of its channels and ditches, honking low, the sentinel ganders with their proud heads and necks showing above the grassy levels around them. One afternoon when I was walking across the coarse cover of the marsh—which seemed to stretch far off like the pampas, with its indefinite sky and a wide-spread travel of birds—I caught sight of a deer running up behind me, some fifty yards away. It was a doe, with a dun-colored winter coat; and seeing me, she swerved suddenly and headed out toward the middle of the marsh. The waters of January are bitterly cold, but the doe swam a wide channel to get to a small island in the middle, and there she stayed, shaking and scratching now and then, stirring around in an area that became more and more circumscribed as the tide began to rise and the waters widened. I left her a couple of hours later in the gathering dusk, a dark, distant little figure, hunched up far out on the marsh. Deer can swim for several miles, even in icy waters, so she undoubtedly swam back after I was safely out of the way, perhaps after dark when the tide started to go down again. Still, I was troubled by what I had caused, and I came back early next morning to reassure myself that she was gone.

When night comes on, the dark flat marsh has a look of absolute secrecy. The cold winter wind completes its isolation. A few last birds may fly up over it, or twist and cry in the wind and then drop down and disappear. What quick movements, starts, flicking actions, what flight there may be left is at last hidden, downed completely, and the wind and surf sounds wash out all else.

There is secrecy and at the same time a desolation in the marsh, the desolation of life pared down to absolute essentials. It offers no luxury but motion in its tidal context, an absolute minimum of redundance. It is a spare unity, even with all its life and light, and the colors that play over it throughout the years, a whole which only accepts those parts which are necessary to it. This marsh is on its own, with ancient standards of simplicity. To find fulfillment in them would be luxury indeed. The lights begin to go on in the houses that stand over its inner shore, as evening advances. A plane drones in the sky. The marsh's flat, wind-blown darkness is alone, and seems to say that all life is received by those bare standards, that we are all helplessly interdependent and obligated to tides that none of us can turn.

XV

The Uses of Light

In the face of what it offers, I have said very little about the great beach. In some respects it is indefinable even as a geographical entity, in spite of the fact that it represents a range of sandy shore line that extends for thousands of miles to the south of it. It fluctuates so, and it is so closely associated with the sea in that respect, that the term "transition zone," while generally appropriate, seems a little misleading. It is made of land materials but it is not exactly a land boundary. Cape Cod, whose Outer Shore it defines, is as narrow and exposed as a spit or shoal by comparison with the continent behind it. In any case, the beach in its grand exposure, its instability, seems closer to the sea than land, and that may be the reason why many visitors, bound to the inland world of human claims, have often expressed the feeling that it looks untouched.

Small white waves on the sea surfaces beyond the beach may scud like birds while surf and sand are resplendent in green and silver; or an evening wind from the north blows over sandbanks and beach grasses, coming on in hesitant rushes, the gray waters conflicting over shark-gray shoals, and clouds standing off over the sea. Sometimes the surf strikes and hisses like snakes curling along the sand. Sometimes it rises up with green-marbled surfaces, roaring and falling with ponderous formality. Beach and sea are always involved in mutual storms and plays of light, mutual readjustments beyond our control.

The beach is naked, malleable, ready to move and be moved. It is invested with the vast balance of the oceanic tides. It is part of the systems of wind and weather. It is a receiving ground for light. For these and countless other reasons it is a power, with an expression made up of all its communicant and communicating energies, their substance, and formality. It sweeps on in a long curving line that is a definition not only of a bound but a horizon, a sea, and a sky. It expresses growth and the stunting of growth, destruction and its holding back, the violent storm, the offshore summer swell, the heat and cold. Many languages, heard or unheard by human ears are in it, and much that is unknown to us. Its long roving ways invite a man to the space in which life is shaped and perpetuated, invite him, in a sense, to where he is unable to go, where nothing is promised; but it is human perception and realization that it brings out, not security, a man's coming at the size of the natural realm with its unceasing winds, where the birds fly in with a grace and concordance that he will find he knows, by virtue of a primal inheritance.

Life has particular, even narrow, definitions, like those distinct levels of the beach to which different species are adapted. Plants and animals that live in the sand, on the cliff tops, or on the ocean floor beyond the surf, have been responding in the same way for millions of years. All this is well known to natural science. In fact, to make too many ignorant and loose generalities about it is probably an offense to the circumstances; but together with precise conditions goes a vast scope, a space, and a speed like the overworld racing of the ocean tide. The beach and its sands, the waves that cut them away or build them up, its long roaming, and its give and take with respect to the sea, involves a balance that cannot be separated from the globe itself, with an age and a future where time is nearly lost. Seaweed, crabs, shells, fish, or birds are all ancient, exact, and well defined. (It may take hundreds of thousands of years to change the shape of a head or a claw.) They are also part of a motion which is not changed into a machine by being called perpetual. In any case, each form, through the countless passages of light and dark, was endowed with a joyful resistance to finality.

Within the shifting landscape of the sea beach all action, each affinity, and each response, seems controlled and at the same time free and exemplary. The elements agree in making the junctures of light unparalleled. Here are the eternal cross-ways of tides, wind, and sunlight, full of an indefinite potentiality that comes more clear to human eyes because of their lack of obstruction. I think of one area in particular which combines this wide range of view with conflict and meeting more than most. Where the great beach has its last break at Chatham, before the long sandy island, or sandspit, of Monomoy, the tide races through and behind it into Chatham Harbor, and toward the west it flows between Morris Island and Monomoy into Nantucket Sound. There is so much intermingling of currents and tides, so many effects of sky light and clouds and direct sunlight spreading over this area, together with sea smells and varying winds, as to give it an effect of constant remaking and realliance. From the Morris Island shore the surf shows up in the distance above the long low barrier of the Outer Beach like a mirage of waves and when the north wind flings back spray on their crests they might be great dolphins plunging forward through the sea. The cloud masses shift and change, tall in the spring or autumn sky, over sand and long stretches of green and blue water.

Morris Island's sandy, wind-punished shores are full of dead oak and pine, the oak still standing in many places stripped of its bark, a slick stonelike gray, and the ground is covered with a tangle of thickets and beach-grass perimeters all leading to a rim of salt-marsh grasses that joins with sandbars and tidal flats beyond. Through spring and summer and during the early fall when the shore birds have not yet migrated, shoals and bars and flats are covered at low or half tide not only with shifting light over shallow waters but a silvery crying. Wind, foghorns, gulls screaming, shore birds piping, sometimes the faint or bell-like notes of inland birds, planes, perhaps an occasional ship's bell heard or imagined, all sound through the seasons.

During the winter the channels provide some shelter and feeding grounds for ducks and of course the gulls station themselves here and fly up at all times. This point where the tides turn a corner is a contrast in force and influence. There is the rolling and tossing of the open ocean not far away; local waters are agitated by the wind, colored by sun and sky, and always running in or out along the shore; there is a tidal rip in one area where currents meet; a great rushing tidal stream at one place, calm, easing waters in another. Within the framework of tides and storms water changes the shape and volume of the sand as it does along the Outer Shore. There is a holding, a circling as the Atlantic waters meet and turn. The earth seems to toss with all their rhythmic interplay. Flying or flying sounds are in the hands of oceanic light and surprise. There is a special tension in things that responds to a great order and sway.

Whatever animals come here to subsist, or migrate through, have an alliance with this energy, a tidal intensity of their own, taking part in all the contrasts and conflicts of the environment. During the late spring for example, you walk from a relative silence on the Morris Island shore to wide breath and sound a hundred yards away. When the birds are nesting leafy tangles and trees collaborate with them in their concealment. Singing has died down. There are only occasional calls from small birds half-hidden in the leaves, flying from one protected spot to another, and now and then the nestlings make squeaky or rasping little cries in the demands of hunger, but just beyond them the sky is open and bright with action, and there is no need to hide.

In spring and summer the terns are in constant bright evidence over the open water and the sands. The woodpeckers and the sparrows stay with trees or grasses; the terns are birds of the ocean airs and long white shores, their complement and grace. Thousands of terns nest at Tern Island, on the shore of Chatham Harbor, and through the summer months and early fall there are always a great many off Morris Island and Monomoy, diving for fish. They are sharply made, lithe fliers with a nervous excitability that is peculiar to them. Flocks will hover over a stretch of tidal water where schools of small fish are running and they will fairly batter the water, making a loud sound like paddle wheels as they cover it with points of spray. Hundreds, crying harshly, hover some five or six feet up, dropping and rising continually. Many of them dip forward with wings folded slightly, but others, a little higher up, make steeper dives, hovering against the wind, their wings beating hard, to drop, twisting slightly, and then dive with wings back and head down, sharply and precisely. I have thought that terns seldom miss when they have a fish in sight, but during this kind of mass fishing, particularly when they dip forward as if to pick the fish up and try again, it does look like a matter of trial and error. Also, depending on the season, there may be a number of immature birds in the flock that are not as skillful as their elders.

The terns are expert performers in every way. They are small and light with strong, angled wings that can carry them over thousands of miles. They have range, persistence, a bright balance that carries them through the mighty and punishing wilderness in which they live. At the same time, that lovely harsh crying excellence in the form of a tern is fragile, even ephemeral. Terns, in the early period of their lives at least, are expendable, like fish. Common terns especially have large breeding colonies that are extremely vulnerable to human encroachment as well as rats, cats, dogs, skunks, and other marauders, and they definitely need protection. Their existence as a race is hazardous under the best of conditions. The sandy islands or peninsulas which they use for nesting sites may be flooded by storm tides in the spring, destroying quantities of eggs or young birds. An adult tern might live to between fifteen and twenty years of age, although their annual mortality is 23 per cent, and their chance of reaching adulthood is fairly slim, tern mortality in the first year averaging about 80 per cent.

The hard statistics make short lives of many species, while the sun and sea keep their steady and infinite relationship. There is a quality of sacrifice in all life. Nothing is spared in its duration, and at the same time in the uniqueness of its making, as the fires burn. The results of evolution may seem haphazard in many respects, and the processes of nature to involve enormous waste; but natural continuity holds all things in high honor, through the fine balances of life and death. The forms of fish or tern, with their own transmutations of energy, are as excellent as they are perishable.

The tides run the channels with an almost sentient, purling calm during the burning days of summer and early fall. They lift into marshy shores and over sandy flats, and then subside. Sometimes the fog comes on in the afternoon and the deep foghorn groans through sheets of silver under the wind, a low curtain moving on and parting slightly here and there, the sunlight showing intermittently. Tiny black snails move over the flats at low tide, some absolutely still, others moving slightly with black antennae protruding and their feet probing forward. Small fish dart in the pools and hover in the tidal currents. There is an over-all mewing, chuckling, and crying, with an occasional "huh, huh" from a gull flying slowly overhead, as the light shifts with the breath of wind over water.

Gray and white ring-billed and herring gulls, occasional laughing gulls with black heads and red legs, and terns, preen, stalk, stand off in the distance, and fly up intermittently. Ringed plovers run hurriedly forward over the sands and through the shallows. A yellowlegs, tall and limber, stalks, bobs, and probes along an edge of the shore. Black-bellied plovers, big-headed, short-billed, stocky by comparison, trot through the waters, standing up straight at intervals, while the yellowlegs suddenly races back and forth on its hunt for food, turning back on its shadow. These shore birds fly off fast when disturbed, crying out, the black-bellied plover with a sweet whistle of its own.

Gentle rising and falling of the tide over ribs of sand; swirling fogs; burning sun with spokes slanting down through clouds over the rim of the world, letting in calm soft lights, green and pink and pearly across sand and rivulets and pools, or cruelly glittering diamonds over the water. Light and water and wings flow in and flow past, the motion of ages, all actions being synchronized, as the hovering and diving of the tern is synchronized with the fish it catches, part of the indefinite combinations of things in a universe of motion. Over these waters and receptive sands life crawls or flies, dives, halts, stops, and starts, wildly, with quick hearts beating, or scarcely a heart at all, blind, or vibrant with sight, probing with accuracy and speed or merely moving at random.

They are all elements in a great exchange—this ardor and play of one instant in time, an instant that is equal in importance to all others. I stand here at the apex of one day. Here out of a thousand years is another advent, another chance for action, another use for sight, in the beautiful agreement of all contrary, separate, and divided things.

I remember one evening at Morris Island in the latter part of August, with the day beginning to fall and the surf's dull roar sounding from the sands of the great beach, a beach behind me, still beyond me, still in a sense not walked. The tide started to ebb, flicking lightly against the shore, lapsing with the evening as if the sea had an easy courtesy of its own, and with the smoky sunset low on the western sky, the waters moved out over gray sands. There was a perfect symmetry to the evening. Terns flew over, light, airy, floating with a swallow's beat, but deep, sure, and strong. Little sanderlings and red-backed sandpipers, half-seen in the dusk, ran through reflections in the shallow waters at the edge of the tide, part of its coolness and flow, the little waves in banked rows rippling. The birds tripped forward and dipped to the mirrored salmon, copper and crystal in these waters, in a communication. The terns trilled harshly and sometimes their bodies trembled as they beat up against the light wind and changed position. A single herring gull stood still on a hummock at the tide's edge like an Indian in a ritualistic acceptance of darkness coming on. The order of change and constancy began to take light's fire and warmth and its colors away, in the graduated motion of the sky, along with all flying elements like the terns, like thought, and the unimagined combinations of being. The wavelets edged out. The sanderlings started to flit off and disappear. Finally there was no turning back the authority of night.